# Semantic Applications

Thomas Hoppe · Bernhard Humm
Anatol Reibold
Editors

# Semantic Applications

## Methodology, Technology, Corporate Use

 Springer Vieweg

*Editors*
Thomas Hoppe
Datenlabor Berlin
Berlin, Germany

Bernhard Humm
Fachbereich Informatik
Hochschule Darmstadt
Darmstadt, Germany

Anatol Reibold
Ontoport UG
Sulzbach, Germany

ISBN 978-3-662-55432-6    ISBN 978-3-662-55433-3 (eBook)
https://doi.org/10.1007/978-3-662-55433-3

Library of Congress Control Number: 2018936133

Printed on acid-free paper

This Springer Vieweg imprint is published by the registered company Springer-Verlag GmbH, DE part of Springer Nature.
The registered company address is: Heidelberger Platz 3, 14197 Berlin, Germany

# Preface by the Editors

"Why are there hardly any publications about semantic applications in corporate use?"

We asked ourselves this question a few years ago. Most publications about semantic technologies and the Semantic Web are primarily centered around the technology itself, just illustrating them either by toy examples or isolated applications. By developing semantic applications for corporate use, we gained some expertise ourselves, and we were highly interested in exchanging and sharing this knowledge with peers and learning from each other. Therefore, we established the Corporate Semantic Web community in Germany.

In the years 2014–2017, we organized annual Dagstuhl workshops. Making our learnings available to the wider community was our intention from the beginning, so we published our first book "Corporate Semantic Web – Wie semantische Anwendungen in Unternehmen Nutzen stiften" (in German) in 2015. Subsequently, we published two articles on emerging trends in Corporate Semantic Web in the Informatik Spektrum magazine. Due to the high interest in those publications, we decided on starting a new book project. We have deliberately chosen English as the language for this book in order to share our experience with a worldwide community. We dedicated the 2017 Dagstuhl workshop to this book because we wanted to create more than a loose collection of articles: a coherent work which demonstrates the various aspects of engineering semantic applications.

Semantic applications are slowly but steadily being adopted by corporations and other kinds of organizations. By *semantic applications* we mean software applications which explicitly or implicitly use the semantics (i.e., the meaning) of a domain in order to improve usability, correctness, and completeness. We would like to show how to develop semantic applications in a broad range of business sectors. This book is a collection of articles describing proven methodologies for developing semantic applications including technological and architectural best practices. It is written by practitioners for practitioners. Our target audience includes software engineers and knowledge engineers, but also managers, lecturers and students. All of our co-authors are experts from industry and academia with experience in developing semantic applications. One result of the intense community effort over the years is an increasingly aligned understanding of do's and don'ts in developing such applications.

Schloss Dagstuhl, "where Computer Science meets", has been a wonderful host and is a true incubator for developing and sharing new insights. Therefore, our first thanks go to the friendly and competent staff members of Schloss Dagstuhl.

The most important contributors to this book are, of course, our co-authors: 45 experts in their fields and also in writing high-quality texts. Thank you very much for the great collaboration and for meeting (most) deadlines! We would like to thank in particular Wolfram Bartussek, Hermann Bense, Ulrich Schade, Melanie Siegel, and Paul Walsh, who supported us as extended editorial board. Especially, we would like to thank Timothy Manning for proofreading the entire book and literally suggesting hundreds of improvements!

Finally, we would like to thank the team at Springer, particularly Hermann Engesser and Dorothea Glaunsinger for supporting us over the last years and Sabine Kathke for guiding us in this book project. The collaboration with you has truly been friendly, constructive, and smooth.

Darmstadt and Berlin, Germany, December 2017
Thomas Hoppe, Bernhard Humm, and Anatol Reibold

# Contents

# List of Contributors

**Wolfram Bartussek**  OntoPort UG, Sulzbach, Germany

**Ulrich Beez**  Hochschule Darmstadt, Darmstadt, Germany

**Christoph Beger**  University of Leipzig, Leipzig, Germany

**Hermann Bense**  textOmatic AG, Dortmund, Germany

**Jürgen Bock**  KUKA Roboter GmbH, Augsburg, Germany

**Michael Dembach**  Fraunhofer-Institut FKIE, Wachtberg, Germany

**Tilman Deuschel**  Hochschule Darmstadt, Darmstadt, Germany

**Kerstin Diwisch**  Intelligent Views GmbH, Darmstadt, Germany

**Felix Engel**  University of Hagen, Hagen, Germany

**Anna Fensel**  University of Innsbruck, Innsbruck, Austria

**Christian Fillies**  Semtation GmbH, Potsdam, Germany

**Benjamin Gernhardt**  University of Hagen, Hagen, Germany

**Georg Grossmann**  University of South Australia, Mawson Lakes, Australia

**Matthias Hemmje**  University of Hagen, Hagen, Germany

**Heinrich Herre**  University of Leipzig, Leipzig, Germany

**Thomas Hoppe**  Datenlabor Berlin, Berlin, Germany

**Jens Hülsmann**  ISRA Surface Vision GmbH, Herten, Germany

**Bernhard G. Humm**  Hochschule Darmstadt, Darmstadt, Germany

**Lukas Kaupp**  Hochschule Darmstadt, Darmstadt, Germany

**Sabrina Kirrane**  Vienna University of Economics and Business, Wien, Germany

**Markus Loeffler**  University of Leipzig, Leipzig, Germany

**Wolfgang Mayer**  University of South Australia, Mawson Lakes, Australia

**Frank A. Meineke**  University of Leipzig, Leipzig, Germany

**Victor Mireles**  Semantic Web Company GmbH, Wien, Austria

**Hesam Ossanloo**  Braunschweig, Germany

**Oleksandra Panasiuk**  University of Innsbruck, Innsbruck, Austria

**Tassilo Pellegrini**  University of Applied Sciences St. Pölten, St. Pölten, Austria

**Anatol Reibold**  OntoPort UG, Sulzbach, Germany

**Katja Rillich**  University of Leipzig, Leipzig, Germany

**Ulrich Schade**  Fraunhofer-Institut FKIE, Wachtberg, Germany

**Fabienne Schumann**  dictaJet Ingenieurgesellschaft mbH, Wiesbaden, Germany

**Matt Selway**  University of South Australia, Mawson Lakes, Australia

**Melanie Siegel**  Hochschule Darmstadt, Darmstadt, Germany

**Jan Stanek**  University of South Australia, Mawson Lakes, Australia

**Simon Steyskal**  Siemens AG Österreich, Wien, Austria

**Henrik Strauß**  Semtation GmbH, Potsdam, Germany

**Markus Stumptner**  University of South Australia, Mawson Lakes, Australia

**Robert Tolksdorf**  Berlin, Germany

**Alexandr Uciteli**  University of Leipzig, Leipzig, Germany

**Tobias Vogel**  University of Hagen, Hagen, Germany

**Paul Walsh**  NSilico Life Science, Dublin, Ireland

**Jason Watkins**  University of Hagen, Hagen, Germany

**Frauke Weichhardt**  Semtation GmbH, Potsdam, Germany

**Rigo Wenning**  European Research Consortium for Informatics and Mathematics (GEIE ERCIM), Sophia Antipolis, France

**Karl Erich Wolff**  Hochschule Darmstadt, Darmstadt, Germany

**Wolfram Bartussek** Dipl.-Inform. Wolfram Bartussek (CTO, OntoPort) studied electrical engineering with an emphasis on control theory at the Technical University of Darmstadt and was one of the first students in Computer Science in Germany at the University of Karlsruhe. He then worked for 3 years as a scientific assistant under Prof. D. L. Parnas focusing on the design and verification of operating systems kernels and on software engineering. After 25 years as CEO of a software house, he returned to research. About 2 years ago he established OntoPort with two other seasoned consultants, now focusing on language technologies and the application of ontologies. Wolfram Bartussek continued lecturing since his time as scientific assistant mainly on software engineering, programming, and software quality management at several German Universities and currently, since 1999, at the University of Applied Sciences at Darmstadt on information systems engineering, knowledge management and language technologies.

**Ulrich Beez** Ulrich Beez is a Research Associate at the Computer Science Department of Hochschule Darmstadt – University of Applied Sciences, Germany. At Hochschule Darmstadt in 2015, he received his master's degree while working on Terminology-Based Retrieval of Medical Publications. At Frankfurt University of Applied Sciences in 2013, he received his bachelor's degree while working on Continuous Integration. Prior to his academic career he was working for a large IT company.

**Christoph Beger**  Christoph Beger works as a scientist, software developer and system administrator. He received the degree of Master of Computer Science with focus on Medical Informatics from the University of Leipzig, Germany. Since 2013 he administers the Growth Network CrescNet in Leipzig, which supports early detection of growth disorders of German children. Since 2016 he participates in the BMBF funded project Leipzig Health Atlas.

**Hermann Bense**  Hermann Bense studied Computer Science at the Technical University of Dortmund and received his diploma in 1980. He founded the internet agency bense.com in 1999 and the text generation company textOmatic AG in 2015. In both companies, he works as CEO and is responsible for software development and research in the fields of ontologies, Artificial Intelligence, databases, content managements systems, search engine optimisation (SEO) and natural language generation (NLG). In the Google DNI sponsored project 3dna. news, Hermann Bense works on the hyper-personalisation of multi-language news based on Big Data from the fields of weather, finance, sports, events and traffic.

**Jürgen Bock**  Dr. Jürgen Bock graduated in 2006 from Ulm University, Germany, as "Diplom-Informatiker" and from Griffith University, Brisbane, Australia, as Bachelor of Information Technology with Honours. Intensifying his research in the area of formal knowledge representation and Semantic Web technologies at the FZI Research Center for Information Technology in Karlsruhe, Germany, he received his PhD in 2012 from the Karlsruhe Institute of Technology (KIT). In 2015 he joined the Corporate Research department of KUKA Roboter GmbH, Augsburg, Germany, where he is coordinating a team of researchers in the area of Smart Data and Infrastructure.

**Michael Dembach** Michael Dembach was born on April 21st, 1987 in Brampton, Canada. He studied German and English language at the University Bonn and received his Master of Arts in 2013. He works for Fraunhofer FKIE since November 2013 and is mainly responsible for ontology engineering and natural language processing.

**Tilman Deuschel** Tilman Deuschel is a PhD student at the Cork Institute of Technology, Ireland. He is exploring how to improve the user experience of adaptive user interfaces, a type of graphical user interface that automatically change their appearance during runtime. He is also co-founder of a software agency that provides tools to support psychotherapy patients and therapists. He is dedicated to user centered design, requirement engineering, usability engineering and user experience design of interactive media.

**Kerstin Diwisch** Kerstin Diwisch received the M.Sc. degree in Information Science from the University of Applied Sciences Darmstadt, Germany, in 2013. Currently, she is a Ph.D. candidate of Matthias Hemmje at the University of Hagen, Germany. She is working on ways to support archives in the Cultural Heritage domain to digitize and link their collections. Aside from her studies, she works as a Software Engineer at Intelligent Views GmbH, mainly in the field of semantic technologies.

**Felix Engel** Dr. Ing. Felix Engel studied applied computer science at the university in Duisburg-Essen, Germany where he received his Diploma in 2009 from the Department of Information Science. Since 2009 he is working for the FernUniversität in Hagen at the chair for Multimedia and Internet Applications. At this chair he successfully defended his PhD thesis in 2015 and has since been engaged as a postdoctoral researcher. Within this engagement, he is involved in different teaching activities and has contributed to various national and international projects in the area of digital preservation and knowledge management. In this context, Felix

Engel has authored and co authored various scientific publications at national and international conferences. Furthermore, he has worked as program committee member, reviewer and organizer of conferences/workshops. His research interests cover knowledge management, digital preservation, semantic technologies and information retrieval.

**Anna Fensel** Dr. Anna Fensel is Senior Assistant Professor at the University of Innsbruck, Austria. Anna has been a doctoral student in Informatics at the University of Innsbruck, where she has defended her PhD thesis in 2006. Prior to that, she has received a diploma in Mathematics and Computer Science equivalent to the master's degree in 2003 from Novosibirsk State University, Russia. Anna has been extensively involved in European and national projects related to Semantic technologies and their applications (in areas such as energy efficiency, smart homes, tourism and telecommunications), as an overall coordinator, a local project manager and a technical contributor. She has been a co-organizer or a Program Committee member of more than 100 scientific events, a reviewer for numerous journals, and a project proposals evaluator for the European Commission, as well as national research funding agencies. She is a (co-)author of ca. 100 refereed scientific publications.

**Christian Fillies** Christian Fillies studied Information Science at the Friedrich Alexander University in Erlangen-Nürnberg and at Technical University of Berlin, graduating from there with a diploma. Early on, he specialised in artificial intelligence, natural language processing and object-oriented systems with applications in projects for simulation of office processes at GMD and FAW Ulm. Following this, he worked for several years in developing tools for business process representation and modelling in Berlin, Germany and in California, USA. In 2001 Christian Fillies was one of the founders of Semtation GmbH, with its product SemTalk[R] focusing on the development of semantic modelling tools on the Microsoft platform and has been the engineering lead and Managing Director since then.

**Benjamin Gernhardt** Benjamin Gernhardt received the M.Sc. degree in Computer Science from the University of Applied Sciences Dresden, Germany, in 2013. Currently, he is a Ph.D. candidate of Matthias Hemmje at the University of Hagen, Germany. I addition to this, he is working as a IT-Manager of a Software Company in Munich, Germany.

**Georg Grossmann** Georg Grossmann is a Senior Lecturer at the University of South Australia. He is working on the integration of business processes and complex data structures for systems interoperability and has applied his knowledge successfully in industry projects. Current research interests include integration of service processes, ontology-driven integration and distributed event-based systems. He is currently Co-Chief Investigator in the Data to Decisions Cooperative Research Centre, Steering Committee Chair of the IEEE Conference on Enterprise Computing (EDOC) and Secretary of the IFIP Working Group 5.8 on Enterprise Interoperability.

**Matthias Hemmje** Matthias Hemmje received the Ph.D. degree in Computer Science from the University of Darmstadt, Germany in 1999. He is Full Professor of Computer Science at the University of Hagen. His teaching and research areas include information retrieval, multimedia databases, virtual environments, information visualization, human-computer interaction and multimedia. He holds the Chair of Multimedia and Internet Applications, at the University of Hagen.

**Heinrich Herre** Heinrich Herre is head of the research group "Ontologies in Medicine in Life Sciences" at the Institute for Medical Informatics, Statistics and Epidemiology (IMISE) and emeritus professor at the Institute of Computer Science of the University of Leipzig. His research interests cover topics in formal logic, applied ontology, and artificial intelligence. He is the founder of the group "Ontologies in Biomedicine and Life Sciences" of the German Society for Computer Science, and board member of several Journals and Series (Applied Ontology, Axiomathes, Categories: de Gruyter). H. Herre teaches Applied Ontology at the University of Leipzig.

**Thomas Hoppe** Thomas Hoppe works as a Data Scientist and Knowledge Engineer. He is a lecturer for information systems, content management, data- and text mining and search technology at the Computer Science Department of the Hochschule für Technik und Wirtschaft Berlin – University of Applied Sciences, Germany. He received his doctorate from the University of Dortmund (now Technical University Dortmund) and the degree Dipl.-Inform. from the Technical University Berlin, Germany. In 2008, together with three associates, he founded Ontonym GmbH for the development and commercialization of semantic applications, which he led as CEO until 2015. In 2014 he founded the Datenlabor Berlin (www.datenlabor.berlin) for the purpose of developing customer specific data products for SMEs and for quality assurance and certification of semantic applications and predictive models.

**Jens Hülsmann** Jens Hülsmann is a Senior Research Scientist at ISRA Surface Vision GmbH, a manufacturer of machine vision solutions from Herten, Germany. He joined the company in 2014 and is in charge of the development infrastructure software for optical inspection systems and image processing algorithms. Prior to this, he worked as a Research Associate for the Smart Embedded Systems Group at the University of Osnabrück. His research was focused on the application of robust classification in industrial environments and on uncertainty modelling. In 2009 he graduated with a Master of Science in physics and computer science.

**Bernhard G. Humm** Bernhard G. Humm is a professor at the Computer Science Department of Hochschule Darmstadt – University of Applied Sciences, Germany. He coordinates the Ph.D. programme and is managing director of the institute of applied informatics, Darmstadt (aiDa). He is running several national and international research projects in cooperation with industry and research organizations and publishes his results regularly. Before re-entering university, he worked in the IT industry for 11 years as a software architect, chief consultant, IT manager and head of the research department of a large software company in Germany.

**Lukas Kaupp** Lukas Kaupp is a Research Associate at the Computer Science Department of Hochschule Darmstadt – University of Applied Sciences, Germany. At Hochschule Darmstadt in 2016, he received his master's degree while working on conception, development and evaluation of a container-based cloud platform with focus on user collaboration. In 2014, he received his bachelor's degree while working on a framework to reverse engineer java-based components from source code using a genetic algorithm. Prior to his academic career, he was working for a large IT consulting company and a web-focused startup.

**Sabrina Kirrane** Dr. Sabrina Kirrane is a postdoctoral researcher at the Vienna University of Economics and Business (WU) and co-director of WU's Privacy and Sustainable Computing Lab. Prior to taking up the position at WU, she was a researcher at the Insight Centre for Data Analytics, Ireland. Sabrina's current research focuses on privacy implications associated with the publishing, interlinking and reasoning over personal information. She is particularly interested in the multidisciplinary nature of privacy preservation that spans the humanities, social sciences, information technology, security and legal domains. Moreover, Sabrina is the technical coordinator of an EU project around building technical solutions to support the upcoming European General Data Protection Regulation (GDPR).

**Markus Loeffler**  Prof. Markus Loeffler is full Professor and Head of the Institute for Medical Informatics, Statistics and Epidemiology (IMISE) at the University of Leipzig. His scientific research covers a wide range of subjects in biometry, biomathematics, clinical trial research, systems biology and computational biology. He is Scientific Director of the Center for Clinical Trials (ZKS), Scientific Director of the Interdisciplinary Center for Bioinformatics (IZBI), Director of the LIFE Research Center for Civilisation Diseases, and PI of the consortium SMITH in the BMBF Medical Informatics Initiative.

**Wolfgang Mayer**  Wolfgang Mayer is a Senior Lecturer at the University of South Australia. His research interests include artificial intelligence and knowledge representation methods and their applications in industrial contexts. He has made contributions to information modelling and natural language knowledge extraction methods, scalable data architectures, causal inference from data, technologies for software systems interoperability, expert systems for process and product customization, and fault diagnosis in software systems.

**Frank A. Meineke**  Frank A. Meineke works as Data Scientist and Project Manager at the Institute of Medical Informatics, Statistics and Epidemiology in Leipzig. He studied applied computer science with a focus on computer linguistics in Koblenz-Landau and graduated in 1994 with a degree (Dipl Inform) in computer science. He worked at the Universities of Cologne and Leipzig in the field of systemic medicine and software development (oncologic treatment planning) and received his doctorate (Dr. rer. med.) in 2007 from the medical faculty of Leipzig for his work on stem cell development. He took over a leading position in an IT-department of a local hospital and returned to the University of Leipzig in 2010, where he now works as IT-Coordinator of the Institute and heads up IT groups in various clinical research projects (IFB Obesity, Leipzig Health Atlas, SMITH).

**Victor Mireles**  Victor Mireles is researcher at Semantic Web Company GmbH, Vienna, Austria. He holds B.Sc. and M.Sc. degrees in computer science from the National Autonomous University of Mexico (UNAM). He has conducted research in Mexico, Germany and Austria, in the fields of Computational Biology, Natural Language Processing, and Matrix Decompositions. His current research focuses on the interface between semantics, machine learning, and automatic reasoning.

**Hesam Ossanloo**  Hesam was born in Iran in 1983. His passion for discovering the unknown ignited the will to get out of Iran to learn more about other cultures, science and humanity. He received his M.Sc. degree in computer science from the Hochschule Darmstadt – University of Applied Sciences where he is also doing his Ph.D. in the field of semantic web. In the morning, he is a senior consultant and at night, a scientist doing research on how to improve the search experience for the end-users. In his Ph.D. he is part of a team, which is developing a semantic application for finding software components called "SoftwareFinder" (www.softwarefinder.org). SoftwareFinder uses a domain-specific ontology for offering semantic functionalities to make the user's life a bit easier.

He is also passionate about working on the frontier of science, which is why he is part of the project "Mission To the Moon" at the PTScientists GmbH (http://ptscientists.com), working on landing a rover on the moon.

**Oleksandra Panasiuk**  Oleksandra Panasiuk is a Ph.D. student at the Semantic Technology Institute Innsbruck, Department of Computer Science, University of Innsbruck, Austria. She holds B.Sc. and M.Sc. degrees in Informatics from Taras Shevchenko National University of Kyiv, Ukraine. Her current research focuses on structure data modelling and semantic annotations. She is a co-chair of the Schema Tourism Working Group by STI International.

**Tassilo Pellegrini** Tassilo Pellegrini is a professor at the Department of Media Economics of the University of Applied Sciences in St. Pölten, Austria. His research interest lies at the intersection of media convergence, technology diffusion and ICT policy. Besides his focus on media economics, he is doing research on the adoption of semantic technologies and the Semantic Web among the media industry. He is member of the International Network for Information Ethics (INIE), the African Network of Information Ethics (ANIE) and the Deutsche Gesellschaft für Publizistik und Kommunikationswissenschaft (DGPUK). He is co-founder of the Semantic Web Company in Vienna, editor and author of several publications in his research areas and conference chair of the annual I-SEMANTICS conference series founded in 2005.

**Anatol Reibold** Anatol Reibold studied mathematics and mechanical engineering at the University of Novosibirsk. Being a passionate mathematician, his main interests focus on semantic technologies, data analysis, data science, optimization, transport logistics, coloured Petri nets, GABEK, DRAKON, and polyductive theory. Currently he works as a supply chain analyst and business mathematician at Raiffeisen Waren-Zentrale Rhein-Main eG. He is also one of the co-founders of OntoPort and their chief data scientist.

**Katja Rillich** Katja Rillich finished her biology studies at the University of Leipzig with the diploma in 2003. In 2008 she finished her doctoral thesis on the subject of light-induced calcium increases in glial cells of the guinea pig retina and continued working as a research associate at the Paul-Flechsig-Institute for brain research (University of Leipzig) until 2010. Since July 2010, she works at the Institute for Medical Informatics, Statistics and Epidemiology of the University of Leipzig and has been involved in several groups and topics.

**Ulrich Schade** Ulrich Schade received the Ph.D. degree in Linguistics from Bielefeld University, Germany, in 1990. He is a senior research scientist with Fraunhofer FKIE where he leads a research unit on information analysis. Additionally, he is associate professor to the Department of English, American, and Celtic Studies, at Rheinische Friedrich-Wilhelms-Universität Bonn.

**Fabienne Schumann** Fabienne Schumann is an Information Manager at dictaJet Ingenieurgesellschaft mbH, a service provider for technical documentation and process optimisation in the field of information management, located in Wiesbaden, Germany. At dictaJet, Fabienne is in charge of the ProDok 4.0 project coordination. In 1994 she graduated from the Hochschule Darmstadt, University of Applied Sciences, Germany, with a diploma degree in Documentation and Information Sciences. Prior to this, she graduated from the Sorbonne University in Paris, France, with a Maîtrise degree in Applied Foreign Languages.

**Matt Selway** Matt Selway is a Research Fellow at the University of South Australia. He is working on the extraction of process models from text and their transformation into executable models for a variety of simulation environments, as well as contributing to interoperability projects in industrial contexts. His research interests include Natural Language Understanding, Knowledge Engineering and integration, and software systems interoperability.

**Melanie Siegel** Melanie Siegel is a Professor for Information Science at Hochschule Darmstadt, with a strong background in language technology. She holds a Ph.D. in linguistics and a Habilitation in linguistics and computational linguistics from Bielefeld University. From 1993 to 2000, she worked in the Verbmobil project on machine translation and was responsible for the semantic analysis of Japanese. Her Ph.D. thesis, which was finished in 1996, concerns mismatches in Japanese to German machine translation. The habilitation thesis, which was finished in 2006, is titled "JACY – A Grammar for Annotating Syntax, Semantics and Pragmatics of Written and

Spoken Japanese for NLP Application Purposes". She participated in research projects at the German Research Center for Artificial Intelligence (DFKI) and Saarland University between 1995 and 2006. From 2006 to 2012, she worked as a computational linguist and head of research and innovation at Acrolinx GmbH in Berlin, in the area of automatic consistency checking of technical documentation. www.melaniesiegel.de

**Jan Stanek** Jan Stanek is a lecturer at the University of South Australia (in Business intelligence and Analytics and Health informatics). He is interested in internal medicine, medical clinical databases, medical process analytics and modelling, interoperability in health (especially in pathology and in general practice) and clinical decision support systems. He led projects on the design and development of clinical databases for genetic data collection, and was a member of the team developing a standard for clinical databases with the Royal College of Pathologists of Australasia. Currently, he is working on projects in pathology process analytics and modelling and on general practice data collection and analytics.

**Simon Steyskal** Simon Steyskal holds master's degrees in Software Engineering & Internet Computing as well as Information & Knowledge Management, both from the Vienna University of Technology (TU) and joined Vienna University of Business and Economics in January 2014 as a Ph.D. candidate under the supervision of Prof. Axel Polleres. He is working as a research scientist for Siemens AG Austria and was involved in a joint research project between Siemens AG and WU. He was a member of the W3C RDF Data Shapes working group and involved in the standardization process of a W3C permissions and obligations expression language carried out by the W3C Permissions and Obligations Expression working group.

**Henrik Strauß** Henrik Strauß holds a master's degree in Business Information Science from the Technische Hochschule Brandenburg, University of Applied Sciences. He is working on projects concerning process modelling tools and the use of ontologies on the Microsoft platform at Semtation GmbH in Potsdam, Germany.

**Markus Stumptner** Markus Stumptner is Professor for Computer Science at the University of South Australia (UniSA) and head of the Knowledge & Software Engineering research group. He is the Director of the Advanced Computing Research Centre (ACRC) at UniSA, and is Research Program Lead for Data Management in the Data to Decisions Collaborative Research Centre (D2D CRC). He has worked on knowledge representation and reasoning in a variety of problem areas such as configuration, design, diagnosis, interoperability, service composition and automated model creation from natural language.

**Robert Tolksdorf** Robert Tolksdorf was a Professor for Computer Science at Freie Universität Berlin. His group, Networked Information Systems, was involved in numerous projects on applications of the Semantic Web from 2002 until 2016. He holds a Dr.-Ing. from Technische Universität where he also received his Diplom-Informatiker after studying Computer Science. He was one of the co-founders of Ontonym GmbH, Berlin. www.robert-tolksdorf.de

**Alexandr Uciteli** Alexandr Uciteli received the degree of Master of Computer Science from the University of Leipzig, Germany in 2008. After that, he worked as a software developer at the Max Planck Institute for Evolutionary Anthropology in Leipzig. Since August 2008, he works as Research Associate at the Institute for Medical Informatics, Statistics and Epidemiology (IMISE) of the University of Leipzig and belongs to the research group "Ontologies in Medicine and Life Sciences". His research is focused, inter alia, on modelling clinical metadata, ontology-based information retrieval and ontology-driven software development.

**Tobias Vogel** Tobias Vogel received the Ph.D. degree in Computer Science from the University of Hagen, Germany in 2012. Currently, he is a post-doctoral researcher at the University of Hagen in the research team of Matthias Hemmje. In addition to this, he is working as Managing Director of a Technology Company in Munich, Germany.

**Paul Walsh** Paul Walsh is Professor of Computer Science in the Cork Institute of Technology (CIT) and a Senior Visiting Research Fellow at the University of Edinburgh where he manages research in medical informatics and bio-informatics. He holds a Ph.D., M.Sc. and B.Sc. Hons in Computer Science from the National University of Ireland and has a long list of publications including a number with outstanding paper awards. He has consulted on a wide spectrum of projects, ranging from startup technology companies to managing projects for global corporations. He is a Science Foundation Ireland Research Fellow and is funded under national and international research pro-grammes, such as the Enterprise Ireland and Horizon 2020, where he oversees research in data analytics, machine learning and high performance computing. He is founder of NSilico Life Science, a software company that provides machine learning driven solutions for global clients.

**Jason Watkins** Jason Watkins received his B.Sc. degree in Automation Technology from the University of Applied Sciences Mannheim, Germany, in 2011. At his first job at Bihl+Wiedemann GmbH, he was designing firmware for fieldbus interfaces of industrial devices as an embedded software developer. In 2014, he joined John Deere as a design engineer developing software for frontloader applications, electro-hydraulic front axle suspensions and electro-hydraulic cabin suspensions. Alongside his professional activities, he started studying a master's program in Electrical Engineering at the University of Hagen. Currently, he is working on his master's thesis in the field of Ontology Matching under supervision of Kerstin Diwisch and Matthias Hemmje.

**Frauke Weichhardt** Frauke Weichhardt holds a Ph.D. in Business Informatics after graduating in Business and Mechanical Engineering from the Technical University of Berlin. She was a lecturer for Business Informatics at TU Berlin and then worked as a consultant in business process design and knowledge management projects before, in 2001, she became one of the founders of Semtation GmbH with its product SemTalk$^{(R)}$, focusing on the development of semantic modelling tools on the Microsoft platform and is working as Managing Director for the company.

**Rigo Wenning** Rigo Wenning is the legal counsel of the European Research Consortium for Informatics and Mathematics (GEIE ERCIM), a joint venture of CWI, Fraunhofer, IIT-CNR, INRIA and ICS-FORTH. GEIE ERCIM is the European host of the World Wide Web Consortium (W3C) where Rigo also acts as legal counsel. He is an attorney at law registered with the Munich Bar Association and affiliated with the law firm Frösner & Stadler in Freising. He studied law in Saarbrücken and Nancy. Apart from legal advice, Rigo does research in the area of privacy and security. He has been involved in the PRIME project, the Primelife project and was technical coordinator of the STREWS project on IT-Security. He is now involved with the SPECIAL project that creates a scalable policy-aware linked data architecture for privacy transparency and compliance for big data. Rigo Wenning is a member of the board of Deutscher EDV-Gerichtstag and also a member of the scientific council of the Leibniz Information Centre for Science and Technology University Library (TIB).

**Karl Erich Wolff** Karl Erich Wolff studied mathematics and physics, received the Ph.D. in 1973 and the habilitation in 1978 in mathematics from Gießen University, Germany, where he lectured from 1974 until 1980 at the Mathematical Institute. From 1980 until his retirement in 2011, he taught mathematics at Darmstadt University of Applied Sciences. Since 1983, he is also a member of the research group "Concept Analysis" in the Department of Mathematics at the university of technology TU Darmstadt. Since 2011, he is president of the Ernst-Schröder-Center for Conceptual Knowledge Processing. He has extended Formal Concept Analysis to Temporal Concept Analysis.

# Introduction to Semantic Applications

1

Wolfram Bartussek, Hermann Bense, Thomas Hoppe,
Bernhard G. Humm, Anatol Reibold, Ulrich Schade,
Melanie Siegel, and Paul Walsh

**Key Statements**

1. Semantic applications today provide benefits to numerous organisations in business sectors such as health care, finance, industry, and the public sector.
2. Semantic applications use the semantics of a domain in order to improve usability, correctness, and completeness.

W. Bartussek (✉) · A. Reibold
OntoPort UG, Sulzbach, Germany
e-mail: w.bartussek@ontoport.de; a.reibold@ontoport.de

H. Bense
textOmatic AG, Dortmund, Germany
e-mail: hermann.bense@textomatic.ag

T. Hoppe
Datenlabor Berlin, Berlin, Germany
e-mail: thomas.hoppe@datenlabor.berlin

B. G. Humm
Hochschule Darmstadt, Darmstadt, Germany
e-mail: bernhard.humm@h-da.de

U. Schade
Fraunhofer-Institut FKIE, Wachtberg, Germany
e-mail: ulrich.schade@fkie.fraunhofer.de

M. Siegel
Hochschule Darmstadt, Darmstadt, Germany
e-mail: melanie.siegel@h-da.de

P. Walsh
NSilico Life Science, Dublin, Ireland
e-mail: paul.walsh@nsilico.com

© Springer-Verlag GmbH Germany, part of Springer Nature 2018
T. Hoppe et al. (eds.), *Semantic Applications*,
https://doi.org/10.1007/978-3-662-55433-3_1

3. Developing semantic applications requires methodological skills, e.g., ontology engineering, quality assurance for ontologies, and licence management.
4. Various technologies are available for implementing semantic applications, e.g., data integration, semantic search, machine learning, and complex event processing.

## 1.1    Introduction

Semantic applications today provide benefits to numerous corporations and other kinds of organisations worldwide. This book describes proven methodologies for developing semantic applications including technological and architectural best practices: from data to applications. The methodologies are backed up by a large number of applications that are in corporate use today.

Figure 1.1 gives overview of the book chapters and which methodologies, technologies, and applications in corporate use they cover.

In this chapter, we give an introduction to semantic applications and provide an overview of the most prominent methodologies, technologies, and applications in corporate use.

## 1.2    Foundations

Since not everybody is acquainted with the terminology in the field of semantic applications, we provide definitions of the most important terms used in this book. These definitions are neither intended to be complete nor perfectly consistent with scientific definitions. They are intended to show the intuition behind some of the major terms.

*Semantics* tries to capture and normalise the relationships between words (respectively terms, phrases, symbols etc.) and their meaning. For example, the word "cancer" can have the meaning of a disease or a zodiac sign. The concrete meaning of a term in a formalisation is usually determined by its context, i.e. the other terms used for its definition and the terms related to it. Such formalisations are often called ontologies.

In the context of semantic applications, an *ontology* is an explicit representation of the semantics of the used terms, usually restricted to a specific application domain [1]. The term *ontology* has been defined as a "formal, explicit specification of a conceptualisation" [2], emphasising that the terms are explicitly chosen on a particular level of granularity, or as a "formal, explicit specification of a shared conceptualisation" [3] additionally emphasising that its purpose is to share the meaning of terms between different stakeholders. For example, an ontology for medicine may define melanoma as a disease, Warfarin as a medication, and the relationship between both indicating that Warfarin can be used for treating melanoma. In fact, Fig. 1.1 shows a simple ontology, specifying applications, business sectors for corporate use, methodologies and technologies, and indicating relationships to chapters of this book.

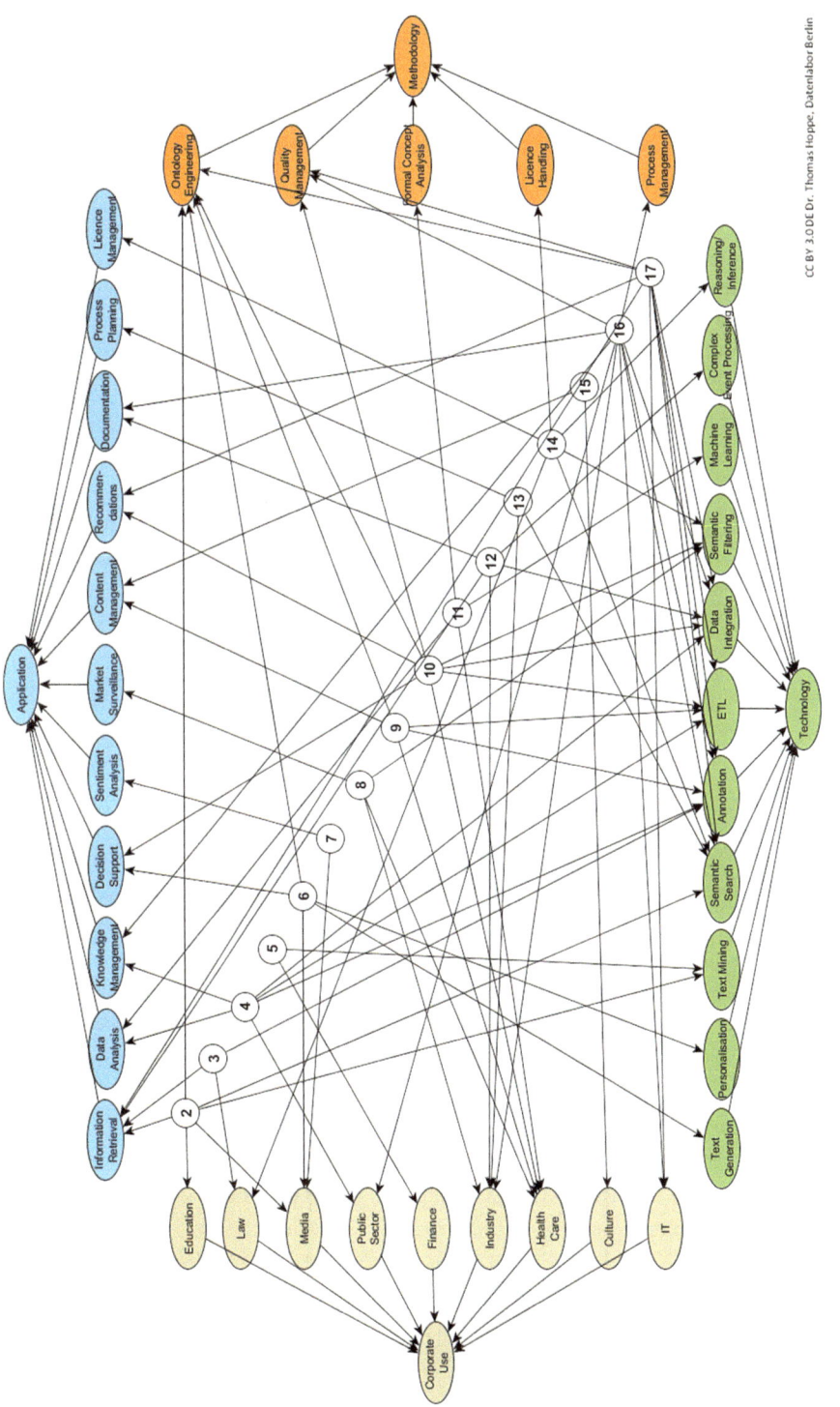

**Fig. 1.1** Overview of the book chapters (White circles indicate chapter numbers. The arrows exiting the white circles indicate the methodologies, technologies, applications and corporate domains covered by each chapter.)

Various forms of ontologies are used under different terms, depending on the complexity of relationships provided. Lassila and McGuiness [4] differentiate between "lightweight ontologies" and "heavyweight ontologies". Lightweight ontologies in particular include controlled vocabularies, thesauri and informal taxonomies. *Controlled vocabularies* are, in their simplest form, just a list of prominent terms used in an application domain, e.g., "melanoma" in the medical domain. *Taxonomies* add hierarchies of broader/narrower terms to controlled vocabularies, e.g., melanoma is a type of cancer, which is a disease. *Thesauri* add additional information to the terms in taxonomies, including prefered names (e.g., "melanoma")*,* synonyms ("malignant melanoma", "cutaneous melanoma*"*), and relations to other terms (e.g. "see also skin cancer"). Heavyweight ontologies extend thesauri by giving the informal hierarchical broader/narrower term relation (i.e. is_a relation) a formal foundation and extending the expressiveness by additional fine grained relations (e.g., gene CRYBG1 is associated with melanoma), definitions (e.g., "melanoma is a malignant neoplasm comprised of melanocytes typically arising in the skin"; Source: National Cancer Institute Thesaurus), properties, and metadata. The focus of ontologies is not only the terminology of a domain, but also the inherent ontological structure, i.e. which objects exist in the application domain, how they can be organised into classes, called *concepts*, and how these classes are defined and related.

*Ontology engineering* is the discipline of building ontologies. It includes methodologies and best practices, e.g., incremental ontology development in tight collaboration with domain experts, and ranges from text analysis of available documents and information sources, over the extraction of information from various data sources, to the modelling of an ontology. *Ontology modelling* covers either the adaptation of existing ontologies, the merging and aligning of several ontologies covering different domain aspects, or modelling the needed ontology from scratch. As a key methodology, ontology engineering is covered in further detail in Sect. 1.4.1.

A *semantic application* is a software application which explicitly or implicitly uses the semantics of a domain. This is to improve usability, validity, and completeness. An example is semantic search, where synonyms and related terms are used for enriching the results of a simple text-based search. Ontologies are the centerpiece of semantic applications.

*Information retrieval* usually subsumes different approaches for obtaining information based on a specific information need from a collection of information sources. Besides pure textual information, it also usually covers image, speech and video retrieval. Nowadays, the most prominent examples of information retrieval applications are general-purpose search engines, such as Google, Yahoo, and Bing. Today, such search engines include semantic search, making them semantic applications. For example, entering "When was JFK born?" in the Google search will result in an information box containing "John F. Kennedy/Date of birth: May 29, 1917". In contrast to general-purpose search engines, domain-specific search applications have a narrower focus but provide more semantic depth. Examples are hotel and travel portals, partner search, job portals, used vehicles websites, etc.

*Data integration* means combining data from different sources, usually in different formats, providing a unified view. In ontology engineering, sometimes various ontologies from the same application domain are integrated in order to improve comprehensiveness. For example, in the medical domain, the National Cancer Institute Thesaurus and the Medical Subject Headings Thesaurus may be integrated to be used together in a semantic application.

## 1.3 Applications and Corporate Use

Adding semantics to information processing might look ambitious or even aloof. As authors of this book, we would like to disagree by presenting real-life examples: Applications that work and that already provide benefits and gains. As the reader, you might like to know whether some of these applications strike your own domain or offer an approach you can exploit for your own work. Therefore, here is a short overview of the applications presented in this book and the business sectors these applications belong to. This overview refers to Fig. 1.1, first taking a look from the "corporate use" point of view (left side in Fig. 1.1) and then from the "application" point of view (top of Fig. 1.1).

### 1.3.1 Corporate Use

The *finance sector* is addressed in Chap. 5, which discusses how to support the preparation, publication, and evaluation of management reports. The *legal sector* is treated in Chaps. 3 and 14. Chapter 3 presents an application that manages compliance questions and problems that arise if open data is exploited. Chapter 14 introduces an application for automated license clearing.

Chapters 12, 13, and 16 present applications addressing the production side of economy, namely the *industry sector*. The application discussed in Chap. 13 supports production planning, in particular where planning and production involves several partners who want to cooperate. Models of production processes or, more generally, business processes can be improved by the use of the automatic annotation tool presented in Chap. 16. The semantic application presented in Chap. 12 helps finding appropriate technical documentation for machinery in error and maintenance situations.

Chapter 2 describes how ontologies can be developed pragmatically in a corporate context. These ontologies serve to improve semantic search, e.g., portals for jobs or continued education (subsumed in Fig. 1.1 under "education"), or within a company's intranet. The application described in Chap. 8 uses an ontology to search the web in order to build corpora of documents about a given topic, which in this case is medical devices.

Semantic applications in the *media sector* include sentiment analysis (Chap. 7) and automatically generating text from structured data, e.g., weather, sporting events and stock reports (Chap. 6).

The *healthcare sector* is handled in various chapters. The application described in Chap. 8 is used for collecting documents about medical devices to support compliance with regulation. Controlled by an ontology it retrieves scientific publications of interest from the Internet, checks them for relevance and organises this information into a searchable repository. Chapter 9 discusses ontology-driven web content development to automatically build a web portal for health informatics. The described method has been used for building the Leipzig Health Atlas, a multifunctional, quality-assured and web-based repository of health-relevant data and methods. Further innovation in the health informatics domain is provided in Chap. 10, which describes an ontology-driven approach for decision support in the field of cancer treatment. Chapter 11 offers a formal mathematical representation for temporal sequences, an approach that is used to provide explanations of terms in broader settings. Applications of this technology in the health domain include gene expression analysis, visualising the behavior of patients in multidimensional spaces based on their genetic data. It has also been used to support the treatment of anorectic patients.

Chapter 15 presents an application that can be used to build cultural heritage archives by matching the vocabulary of different cultural projects. The application discussed in Chap. 4 is assigned to the *public sector*. It supports locating and connecting required data in "data lakes" and the incorporated task of overcoming big data's variety problem.

Finally, the applications presented in Chaps. 16 and 17 can be subsumed under *IT sector*. Chapter 16 describes how semantic technologies can be used to improve the modelling and processing of business processes in order to deliver documents relevant to the current process step at the right time. Chapter 17 is for one's personal or professional use: it supports finding appropriate and required software.

### 1.3.2 Applications

As a reader of this book, you might not find semantic applications you are looking for applied to your own domain. However, applications applied to other domains might also present ideas and approaches you can exploit for your own work. Accordingly, here is the overview from the application point of view.

Getting the right information at the right time at the right place is one of the demands of the digitalisation process, a demand that needs to be supported by *information retrieval* applications. In Chaps. 2, 3, 6, 8, 12, 13, 16, and 17 it is shown how semantics helps to make such applications smarter so that the needed information is no longer buried under piles of barely relevant information. Chapter 4 covers the overlap between *information retrieval*, *data analysis*, and *knowledge management* by showing how these processes can be supported by semantic metadata, in particular if the data to be analysed is highly heterogeneous (variety aspect of big data). Chapter 16 supports the retrieval of process-relevant documents taking the context of the the current process step into account. Chapters 11 and 7 add to the data analysis category whereupon Chap. 11 focus on temporal aspects and Chap. 7 on *sentiment analysis*.

Semantic applications for knowledge management and content management are presented in Chap. 4, tackling the variety problem of big data. Chapter 16 focuses on annotating models of business processes semantically. Chapter 9 describes the development of an ontology-based extension for a Content Management System (CMS) and Chap. 15 discusses how to build archives for cultural heritage. *Documentation* is another facet of information management, so with this in mind, Chap. 12 presents an application for the fast identification of relevant technical documentation for machinery in fault and maintenance situations. Chapter 16 focuses on the documentation of business process models.

An application that supports *market surveillance* and collects scientific documents about medical devices is treated in Chap. 8. *Process planning* is handled in Chap. 13 and *Licence Management* in Chap. 14. *Recommendation* and *Decision Support* are handled in Chap. 10, while Chap. 17 describes a semantic application which gives recommendations about software components to procure in a software development project.

## 1.4 Methodology

The main advantage of knowledge-based systems lies in the separation of knowledge and processing. The knowledge needed is usually represented by ontologies. Ontologies can be classified into so called "top ontologies" representing knowledge common to a number of domains, "domain ontologies" describing the particularities of a domain, and "task ontologies" connecting the domain ontology to an application. Top ontologies are usually developed within research projects. Domain ontologies usually cover the terminology and the informational structure of an application domain for a number of different applications.

The major question for the application of semantic technologies is of course: "From where to obtain the domain ontology?" For certain important domains, extensive ontologies are available (such as medicine and technology), but for specialised business sectors or corporations, fitting ontologies often do not yet exist and need to be created.

Although there exists a number of ontology engineering methodologies, they often were developed in an academic context and seldom tested in the context of real-world or corporate applications. Within this book, a number of different practical approaches are described for engineering the required domain ontologies, which have shown their utility for real applications.

### 1.4.1 Ontology Engineering

If domain ontologies are available for a particular application, it would of course be wasted effort to model the needed ontology each time from scratch. However, often the available ontologies do not completely cover the required application domain. As Chap. 10 emphasises on the example of a personalised health record application, no single ontology in medicine contains all relevant terms and no set of medical ontologies covers all terms of the needed concepts. In this situation, ontology mapping, alignment and integration become the main ontology engineering tasks. Chapter 10 describes one particular instance of ontology engineering, comprising of the transformation of different data formats, the mapping of different semantic fields, cleansing and filtering their content, identifying and handling of duplicates and merging the information into a single target data format.

Sometimes one can find for a particular application domain, such as information technology, some rudimentary classification system, which allows the derivation of the "upper part" of a domain ontology automatically, but which is not as detailed as required by the application at hand. Chapter 17 takes a related, but simpler approach. A light-weighted core of a thesaurus is built by initial identification of a number of "semantic categories" and assigning the domain terminology derived from textual descriptions of software and software components to these categories. Acronyms and synonyms are then related to these terms, and additional tools are used to identify functional synonyms, i.e., terms which are closely related and are used synonymously.

Chapter 2 describes the extension of this approach to a pragmatical ontology modelling approach suited for corporate settings. By means of simple text analysis and the derivation of keyword lists, ontology engineers are guided to first determine the important concepts for an application, called "categorical concepts". These categorical concepts are then used to pre-qualify the extracted keywords in a spreadsheet, giving an ontology engineer hints about the terms to model first and allowing them to track her/his work. Especially for search applications, the modelling of a thesaurus is often sufficient. By establishing some modelling guidelines, it can be ensured that this thesaurus could be transformed later to a more expressive ontology.

Chapter 9 describes a processing pipeline to extend a common CMS, in this case Drupal, by the functionality for automated import of ontologies. The starting point here is the modelling of a domain in a spreadsheet template, converting it to an ontology, optionally optimisation of this ontology by a knowledge engineer and its importing into Drupal's own database. Thus, allowing non-experts to model the content of a web portal with common tools and enriching its knowledge structure before feeding it into the CMS.

Sometimes the structure of an ontology and its representation is constrained by the application making use of it. Therefore, the ontology engineering process needs to respect application-dependent requirements. Chapter 2 describes for example that synonyms of preferred concepts are explicitly marked in order to identify them easily without the need to perform logical inferences and to build up fast lookup tables speeding up the automatic annotation process. In Chap. 6, an RDF-triple store realised with a conventional database is used to gain quick access to the ontology during the text generation process. This requirement imposes a constraint on the ontology design.

Dependent on the domain and application, quality assurance is a topic which needs to be considered during ontology engineering.

## 1.4.2  Quality Management

If we stick to the definition of the term *ontology* as "formal, explicit specification of a shared conceptualisation" [3], the sharing of the conceptualisation implies that different stakeholders agree with the meaning of the represented information. This assures a certain degree of quality of the ontology.

While Chap. 2 shows how the quality of an ontology can be ensured during the incremental modelling by the four-eyes-principle within the ontology engineering process, Chaps. 10, 16, and 17 address the management of information quality by linking information and putting it into the domain context.

The quality of information is particularly important in the medical field. Chapter 10 describes high-quality information as the basis for personalised medicine. Personalisation, i.e., the adaptation of medical treatment to the individuality of the patient, has considerable advantages for patients. Semantic technologies help to link patient data with information in the medical knowledge database. This allows the treatment to be adapted to the specific characteristics of the patient.

Chapter 16 shows how the quality of the information found within organisations is improved by placing it in the semantic context of the business process. In a "filtering bubble", the process models are semantically enriched with roles, documents and other relevant data. The information is modularised and thus reused, which increases consistency. Semantic concepts reduce the information overflow and focus the information found on the relevant context.

In the area of searching for software, the enrichment of data with semantic information increases the quality of search results, as shown in Chap. 17. In a certain context, suitable software is suggested to the user. "Suitable" here means that the software is selected according to features such as license, community support, programming language or operating system. Furthermore, the user is offered software similar to that which they already use.

## 1.5  Technology

After describing applications, corporate use and methodologies, the questions remains open: Which technologies to use? Clearly, this question is dependent on the particular application. The following sections give an overview of some important technologies covered in the remaining chapters of the book.

## 1.5.1  Semantic Search

The term *semantic search* is ambiguous and used by different stakeholders with different meanings. In its most general meaning it summarises any information retrieval technology which uses background knowledge based on some kind of formal semantics, such as taxonomies, thesauri, ontologies or knowledge graphs. In a narrower sense, the term semantic search summarises information retrieval based on a semantified form of keyword search, often some semantic extension implemented on top of Apache Lucene, Apache Solr or ElasticSearch. This narrow sense differentiates it from faceted browsing, from question answering systems, where queries are posed in natural language, and from RDF-based retrieval systems [5], where a clean specification of the sought information is formulated in some SQL-structured query language, such as SparQL or GraphQL. This latter form of retrieval often requires the formulation of queries by technical experts and builds the technological base in Chaps. 13 and 14.

For semantic search in the narrower sense, the application has to cope with rather vague formulated queries, often only expressed by keywords, formulated in the language of the user, which does not necessarily match the language used in the documents or by a corporation. As emphasised in Chap. 2, this kind of semantic search, as for many information retrieval systems, often requires translation between the language of users and of document authors. This translation makes use of some kind of controlled vocabulary defined by the domain ontology, which is used as a bridge to translate search query terms into terms used by information providers.

Chapters 2 and 17 enrich keyword queries by mapping synonyms, acronyms, functional synonyms and hidden terms capturing common misspellings of terms to the controlled vocabulary of a domain ontology in order to augment the search results by information using such terms or to answer corresponding search queries. Chapter 17 additionally shows how term completions and faceted search can be augmented by the used domain ontologies in order to semantify the user interface and the user experience.

Chapter 16 describes in the context of process models how contextual information about the current process step can guide the semantic search for documents as an additional constraint, thus delivering the user the right information at the right time in the right place.

### 1.5.2 Data Integration and ETL

Semantic applications often require *data integration*. Chapter 10 describes approaches for integrating various ontologies for healthcare, since no single ontology has been proven sufficient for the semantic application under development. Chapter 12 shows how to integrate a domain-specific ontology with business data: machine data from the factory floor on the one side and technical documentation on the other side. Chapter 17 outlines how an ontology may be enhanced by data gathered from web crawling and how such an ontology can be integrated with meta data, in this case for software components. Chapter 4 discusses how ontologies and metadata catalogues can aid exploration of heterogeneous data lakes, and simplify integration of multiple data sets.

An established data integration technology is *Extraction, Transformation, and Loading (ETL)*. It is a common approach in business intelligence and data science to transform, augment and analyse information. *Semantic ETL* extends traditional ETL by the process steps: semantic matching, cleansing, filtering, and duplicate handling. As shown in Chap. 8, this process requires tight feedback for continuous improvement. Semantic ETL can also be used within the ontology engineering phase in order to combine the content of different information sources, such as databases, taxonomies, thesauri, ontologies, or knowledge graphs, into an integrated ontology. Chapter 9 explains how available tools such as spreadsheets and specialised translators need to be combined to implement data integration via semantic ETL. Chapter 10 shows that for mapping and aligning ontologies, the necessary processing steps are similar to the steps used in conventional business intelligence and data science tasks.

### 1.5.3 Annotation

The cornerstone for the application of semantic technologies is of course the content; sometimes annotated by the user either in the form of *tags* or *markup*, or sometimes annotated automatically by text analytics. Background knowledge in the form of taxonomies, thesauri or ontologies, can be used to automatically enrich the annotations by related terms. These enriched annotations are sometimes called *fingerprints,* sometimes they are called *footprints;* thus indicating that they characterise and can be used to identify the content they annotate. In a certain sense this enrichment process is also a kind of data integration, since the keywords accompanying the content or derived from them are integrated with the modelled knowledge to form a new knowledge source.

From a technological viewpoint, such annotations allow for a simplification and speed-up of the retrieval of sought information, since they shift part of the semantic analytics from runtime to the indexing phase of the information. From the application viewpoint, the enriched annotations themselves can be objects of data analysis.

The relation between annotation and content can be described by an annotation model as described in Chap. 3 and are attached as any other kind of metadata to the content they annotate. Especially for the retrieval of content from data lakes, annotations become a key consideration with respect to designing and maintaining data lakes, as argued in Chap. 4. Chapter 9 shows how web content can be annotated by concepts from various ontologies.

Annotations are used for filtering of search results and for the annotation of business process models by business objects in Chap. 16. Although not mentioned explicitly, the "tags" used in Chap. 17 are just annotations derived by normalisation directly from the user-supplied tags in order to derive the domain ontology and to describe the software components.

## 1.6 Conclusion

Semantic applications today provide benefits to numerous organisations in business sectors such as health care, finance, industry, and the public sector. Developing semantic applications requires methodological skills, including ontology engineering, quality assurance for ontologies, and licence management. Implementation of semantic applications is often aided by the software engineers having proficiency in current technologies, e.g., data integration, semantic search, machine learning, and complex event processing.

The following chapters of this book provide insights into methodologies and technologies for semantic applications which have been proven useful in corporate practice.

## References

1. Busse J, Humm B, Lübbert C, Moelter F, Reibold A, Rewald M, Schlüter V, Seiler B, Tegtmeier E, Zeh T (2015) Actually, what does "ontology" mean? A term coined by philosophy in the light of different scientific disciplines. Journal of Computing and Information Technology CIT 23(1):29–41. https://doi.org/10.2498/cit.1002508
2. Gruber TR (1993) A translation approach to portable ontologies. Knowledge Acquisition 5(2):199–220. Academic Press
3. Studer R, Benjamins VR, Fensel D (1998) Knowledge engineering: principles and methods. Data and Knowledge Engineering 25:161–197
4. Lassila O, McGuiness DL (2001) The role of frame-based representation on the semantic web. Knowledge Systems Laboratory Report KSL-01-02. Stanford University Press, Palo Alto
5. Dengel A (ed) (2012) Semantische Technologien. Spektrum Akademischer Verlag, Heidelberg

# Guide for Pragmatical Modelling of Ontologies in Corporate Settings

**2**

Thomas Hoppe and Robert Tolksdorf

**Key Statements**

1. Identify the most important classes of terms of your application. Competence queries may help you.
2. If no adoptable knowledge source fits your purpose, use NLP tools to derive keyword lists from documents or search query logs.
3. Inspect these keyword lists for terms belonging to the previously identified categories and mark the corresponding terms as useful, questionable or ignorable.
4. Start to model a thesaurus with these terms first, in order to implement a first proof of concept and derive an initial domain model.
5. Model the thesaurus under guidelines which facilitate a later transition into a full-fledged ontology.
6. Continuously measure your modelling efforts for future tasks, the development of a business case and reassuring of the management.
7. Depending on your application, judge the consequences of defects perceived by the user, how to cope with imperfect language use and spelling of users and how an incomplete model can be continuously refined over time.

T. Hoppe (✉)
Datenlabor Berlin, Berlin, Germany
e-mail: thomas.hoppe@datenlabor.berlin

R. Tolksdorf
Berlin, Germany
e-mail: mail@robert-tolksdorf.de

© Springer-Verlag GmbH Germany, part of Springer Nature 2018
T. Hoppe et al. (eds.), *Semantic Applications*,
https://doi.org/10.1007/978-3-662-55433-3_2

## 2.1    Background

In 2008 we founded Ontonym, a company for developing and commercializing semantic technologies. Initially we focused on the field of human resources (HR) with the intent to innovate job search. The business was quite clear: if we could make job search easier for job seekers, they could more easily find a job better fitting their skills, and if we could help employers to find better fitting candidates, they could fill vacant positions more quickly with better candidates. Obviously, we could create a win-win situation for both. Since both we and our partners had a strong background in semantic technologies and search engine development, we quickly came to the insight that semantic technologies would be ideal, since matching candidates to job profiles is a background knowledge intense application requiring to cope with a rich, ambiguous and somewhat vague terminology, and it would have the potential to become a killer application.

A first demo of how such a matching application could work was developed by the research project "Wissensnetze" at the Freie Universität Berlin and the Humboldt Universität zu Berlin, funded by the German Federal Ministry of Education and Research. The overall goal of the project was to experiment with Semantic Web technologies, which were at that time newly standardized. Amongst various application domains, HR was chosen since a measurable business value could be expected from applying the technologies in that field.

The first step for us was the creation of a demo system for demonstrating the potential of semantic technologies. Since the initial demo application from the research project was not suited for this task, we focused on the development of a semantic search engine for job advertisements. Between 2008 and 2013 we not only developed this search engine, but also adapted and installed it for a couple of customers, extended it to the area of continued education, adapted the base technology for semantic filtering [1] and developed an integrated ontology-based thesaurus for the areas of job advertisements, HR and continued education. This thesaurus, which is still under maintenance, covered about 14,070 concepts, 5,000 equivalent synonyms, 21,570 multilingual terms, 36,900 automatically generated spelling variants, 28,040 subclass axioms and 8,210 relations between concepts by the end of September 2017.

During this period, we developed a pragmatical approach for modelling domain ontologies, which will be described in this chapter and which turned out to work well for corporate search applications. Before we explain our approach, compare it with other existing development methodologies for ontologies and report our experiences, we initially point out some framework conditions we encounter in commercial settings for semantic applications.

## 2.2    Requirements of Corporate Settings

The application context determines what kind of ontology is needed. For example, if the ontology needs to be used as the basis for drawing valid logical conclusions or technical diagnosis; it needs to be sound and consistent. If the ontology is applied in a context where vagueness plays a major role (like medical diagnosis) and some larger number of cases is available, soundness and consistency become less important, since statistics may substitute crisp logic [2, 3]. If the ontology is used for recommendations or comparisons of user interests with available items (e.g. matching of skills and job descriptions, or user requirements and product descriptions), it needs to account for the language ambiguity and the current language usage of the user and the information suppliers. If the ontology is to help users to find what they want, it needs to map their language usage into some kind of controlled vocabulary in order to bridge the language gap between users and authors [4].

Besides these application dependent requirements and the requirements imposed by the chosen development approach, modelling of thesauri and ontologies in a commercial setting needs to respect requirements never spelled out explicitly.

A new not yet widely established technology needs to show its benefits first. Thus, the management often hesitates initially to set up a costly development process with a new technology. First some proof of concept needs to validate the new technology's utility. Independent of what the outcome of such a proof of concept is, it should yield some result, even in the worst case that the technology turns out not to be applicable or useful. Artefacts developed during the proof of concept should be recyclable in order to save at least some part of the investment. In the best case of course, the result of the proof of concept should be ready for further development.

Obviously, a proof of concept should be finished in reasonable time. Thus, not too much effort can be invested into the initial modelling process. A complete and theoretically sound analysis of a domain and the development of a sophisticated ontology can thus not be afforded. Instead, the modelled ontology needs to be available early in order to gain experiences with it and the technology using it. This implies that an initial developed ontology needs to be easily adaptable, so that it can be extended later.

For semantic search applications in corporate settings we thus identified the following requirements:

- Support of the user during the search process
- Mapping the language use of the users into the language used by the authors
- Early availability of the ontology
- Validation of the ontology in a productive setting
- Incremental, data driven extension of the ontology based on users' search queries
- Option to develop a full-fledged ontology out of a simple initial thesaurus

## 2.3  Development Process

Pragmatic modelling of ontologies as we describe it is an inherently incremental, evolutionary approach, which places an ontology early into a production environment in order to validate it and acquire additional data. In order to avoid early investments into the development of a full-fledged ontology whose utility cannot be shown in advance, pragmatic modelling starts with the modelling of a thesaurus which can be refined to a full-fledged ontology if needed. The pragmatic modelling approach we used can thus be characterized as data-driven, incremental, evolutionary, inside-out modelling.

Ontology modelling is usually characterized either as top-down or bottom-up modelling. While top-down approaches start the modelling process by analyzing the domain and differentiating the concepts starting from the most general concepts down to more specialized ones, bottom-up approaches aggregate concepts to more general concepts on demand, if the need arises to subsume concepts.

Our approach starts from a small core of the most important concepts of the intended application (we call them *categorical concepts*) and focuses primarily on the modelling of these concepts, modelling them as appropriate either top-down or bottom-up. This ensures that the knowledge engineer has a clear guideline which terms to integrate into the ontology. By focusing on a clearly defined set of concept categories, the modelling effort can not only be restricted and focused, but also the effort to get a first initial model can be limited.

Our approach is evolutionary in the sense that such a model restricted to important concept categories can be evolved from an initial core by adding additional layers of concepts and connecting them horizontally from the core, thus proceeding from the inside to the outside.

The incrementality of this approach not only results from this style of modelling which extends the available core of concepts by additional layers either vertically or horizontally, but also since it extends the ontology in concept chunks, where each chunk consists of a set of not yet modelled terms identified from new documents or search queries. Hence, the extension of an ontology with our approach becomes purely data-driven by the terms needed to model the reality and their usage in its application domain. Therefore, it avoids wasting modelling effort on concepts which are irrelevant for the application and in the application context.

This approach also places a requirement on its usage: the courage for gaps and imperfection. Since an incremental development of an ontology for a sufficiently complex domain (like medicine, technology or language usage) is inherently never complete and not error-free, it can be applied only for applications which are error-tolerant, where errors are not directly perceptible and where the organization or its management can live with such imperfections.

In the following we describe in more detail the steps of our modelling process.

### 2.3.1   Initial Phase

The initial phase of the modelling process can be considered as a bootstrapping phase, where we need to figure out which concepts to model, from where to obtain a set of initial terms and where we initiate the modelling process.

### 2.3.1.1 Identification of Categorical Concepts

The first step of our approach consists of the identification of the central *categorical concepts* which are important for the intended application.

Consider, for example, the application domain of job search. Obviously, occupations, job titles and functional roles are the most important terms for a job search, followed by business sectors, the required skills and competences an applicant should have, followed by the responsibilities the job holder will have and the tasks s/he works on, places where s/he will work and maybe company names. Although other terms describing the type of employment, working time, benefits, job dependent technical terms etc. also play some role, occupations, job titles, branches, skills, competences and tasks build the most important term categories from the viewpoint of a job offer and job search.

Or, take as another example the application domain of news. Obviously, every news is triggered by some event. Thus, terms denoting events represent the most important term class for describing news, followed by activities, persons, organizations, locations and times. All other terms appearing in news are less important from the viewpoint of novelty.

Consider as another example diagnostic applications. For technical systems the important categorical concepts which need to be initially modelled are components, functions, subsystems, properties, causes and effects. In Chap. 12 these are for example symptoms, causes, and solutions build the categorical concepts. For medical diagnosis the most important categorical concepts are diseases, symptoms, anatomy, medical devices and diagnostic procedures. However, if the application domain of a corresponding medical application is extended from pure diagnostic to treatments, additional categorical concepts become relevant, such as treatments, drugs, substances and side-effects. In Chap. 10 e.g., six different categorical concepts were identified for a melanoma application: medication, activity, symptom, disease, gene, and anatomy.

In Sect. 17.3 the semantic categories "development", "infrastructure", "business", "entertainment" and others were identified as categorical concepts for software component search and recommendation.

### 2.3.1.2 Initial Document Analysis

Having identified the initial set of concept categories, of course the question arises, how do we know which concepts we need to model? Or more precisely, what are the designators for these concepts and how and from where can they be obtained?

Various application fields have a history of effort in harmonizing terminologies to homogenize communication. An example is medicine with existing thesauri, ontologies or databases, e.g. MeSH, SnoMed, UMLS, etc. There is a good chance to find such a knowledge source in the form of vocabularies, standards and others, at least as a starting point, with terms on which already some joint understanding exists in an application field of interest. So a first step should be an extensive research and selection of such sources.

Unfortunately, such information sources do not exist for every application domain. For example, although we can get some categorization of occupations (e.g. *Klassifikation der*

*Berufe* from the German Bundesagentur für Arbeit, or the *International Standard Classification of Occupations* (ISCO)), the description of technical components (e.g. *Thesaurus Technik und Management* (TEMA) of WTI Frankfurt) and product categories (e.g. *Google Product Categories*), there are no such sources for job titles, skills, competences, tasks, news events, types of points of interests, etc. As argued in Sect. 17.3, even in computer science, the existing classification schemes are insufficient to build a semantic search for software components. In such cases we can only start with some analysis of existing documents. Although it is helpful to have a linguistic tool available for extracting noun-phrases from text documents (like the tools described in Chap. 5), it helps in a first step to derive a frequency distribution of nouns from the available documents, order them according to decreasing frequency and to identify terms denoting important categorical concepts.

Sometimes previously posed search queries of search applications can be used to identify categorical concepts using the corresponding terms as input for the modelling process. Not only can the importance of terms be judged by the number of times the same search query was posed, but also it reflects the language usage of the users. The excerpt in Fig. 2.1 shows an example of such a keyword list.

### 2.3.1.3 Inspection of Extracted Terms

Inspecting this list and prequalifying terms of categorical concepts can be done by manual respectively intellectual inspection. For the prequalification, the identified concepts are classified under three broad categories: important (+), ignore (×) and unclear (?), as shown in the 'Classification' column of Fig. 2.1.

This inspection serves the purpose to prioritize and quickly identify important terms which should be modelled. Although it may happen that some important concepts are not identified in the first pass, they may either pop up in some subsequent modelling step or during following modelling phases. Hence, this modelling approach requires an error-tolerant setting.

Column 'modelled' is used to document which terms finally got modelled or which were deferred to a later modelling time point.

### 2.3.1.4 Modelling Process

Our modelling process is based on keyword lists such as those identified using the previously outlined steps. They contain the pre-categorized terms which help to focus the modelling by hiding terms to be ignored, they help to track the process and can be used to record additional information such as the start/end time of modelling, the terms actually modelled, questions or comments.

Modelling starts with the most frequent terms of the keyword list and proceeds downwards. If the term to be modelled is known to or interpretable by the knowledge engineer, s/he can model it directly, otherwise the knowledge engineer first needs to identify a suitable definition of that term. Figure 2.2 summarizes the steps of the modelling process. Green nodes designate modelling tasks, orange nodes mark actions on the keyword list and blue nodes mark transitions between the research and the modelling task.

| | A | B | C | D |
|---|---|---|---|---|
| | Search query | Frequency | Classification | Modeled |
| | daf | 116 | ? | + |
| | Lohnabrechnung | 115 | + | + |
| | arbeits Sicherheit | 113 | x | |
| | Buchhaltungsfachkraft | 113 | + | + |
| | staplerschein | 113 | + | + |
| | english | 112 | + | + |
| | Gesprächsführung | 108 | + | + |
| | Theaterpädagogik | 108 | + | + |
| | Sozialer Bereich | 106 | ? | |
| | IHK | 105 | x | |
| | Steuern | 104 | x | |
| | ada | 103 | ? | + |
| | büroassistenz | 103 | + | ? |
| | html | 103 | + | + |
| | konflikt | 102 | x | |
| | NLP | 102 | ? | + |
| | Medizinisch | 102 | x | |
| | Hygiene | 101 | + | ? |
| | Datenschutz | 98 | + | + |
| | Mediationsausbildung | 98 | + | + |
| | interkulturelle Kommunikation | 96 | + | + |

**Fig. 2.1** Excerpt from a term frequency distribution of German search queries

The following strategy of finding useful definitions turned out to be effective for us. If we need to model an unknown term or an ambiguous term with multiple interpretations, we perform a quick Internet search via Google. If the search results point to some definition contained in a norm, standard or other normative information source, we use this definition. If an entry to Wikipedia can be found, we use the initial part of the corresponding Wikipedia page (whose first paragraph often contains some clear definition) for the interpretation of the term. Dependent on the definition, additional important terms are sometimes mentioned. According to such a definition and the ontology's objective, we model the term and other terms occurring in the definition which appear relevant for the application, such as synonyms, super or sub concepts, abbreviations and translations.

In the early stages of modelling a thesaurus or an ontology, it is quite easy for a knowledge engineer to maintain an overview of the model and to find the place for a new term quickly. However, it can quickly become difficult to maintain the overview, especially if the modelling is done collaboratively by several persons. At the latest, after a couple of thousand terms it is nearly impossible to easily find the right place for a new concept. Thus, in order to identify the right position for a term, the modelling tool should support search within the concepts and labels: first, in order to identify whether the term was

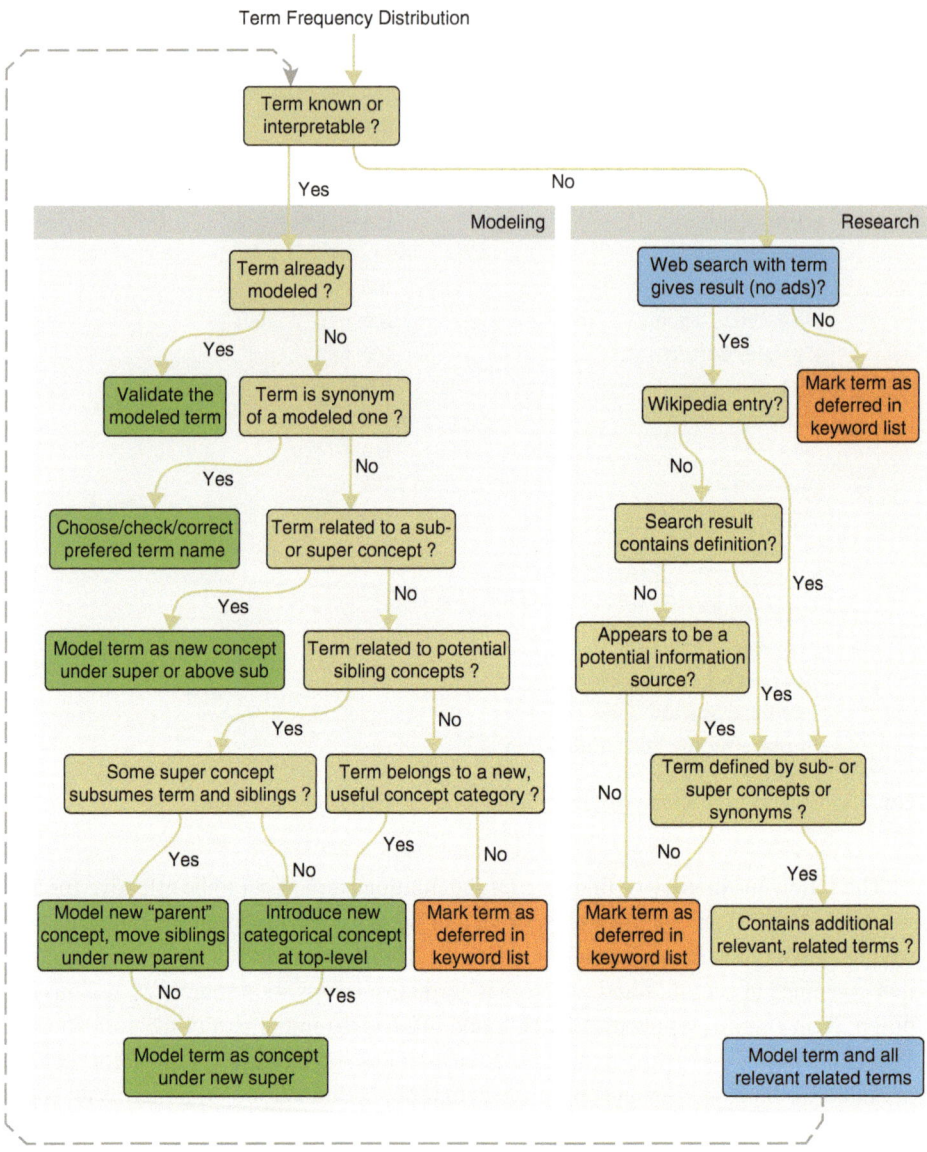

**Fig. 2.2** Workflow of the modelling process

previously already modelled and to validate and correct it if needed, and second to identify concepts which might be related to the term to be modelled.

If no related term or concept can be found within the modelled thesaurus or ontology, the knowledge engineer needs to find the right position within the model. This requires thinking about the meaning of the term. Posing the following questions in unclear cases helped us to identify the proper position of a term:

- What could be concepts subsuming the term?
- How would others name these subsuming concepts?
- How would I paraphrase them?
- Into which category of concepts does the term fall?
- Is the set of objects summarized by the term a superset of some other concept?

This modelling process proceeds down the list to the less frequently occurring terms until a temporal limit, a pre-specified number of terms or some threshold frequency of terms is reached. This helps to limit the modelling effort and to make it calculable.

The developed model has a defined state and can be used for an initial evaluation in the context of a proof of concept or in the production environment. If it was developed based on the term frequencies of the documents, the model fits to the documents and hence an evaluation can be performed on the model and how it supports the document retrieval.

Obviously, the completeness of such an initially derived ontology depends on the representativity and completeness of the underlying text documents respectively on the number of search queries posed within some fixed time interval.

## 2.3.2 Subsequent Phases

Once the proof of concept has resulted in a positive decision to continue with the new technology, the extension and refinement of the modelled ontology become a primary task. This requires the extension of the vocabulary and validation of the previously modelled concepts. While the extension of the vocabulary can follow the process described above, the modelling process needs to be slightly modified in order to validate previously modelled concepts.

Obviously, any modelling is done on the basis of the current understanding and knowledge about the domain, the concepts already available in the ontology and the form or knowledge on the day of the person doing modelling. Hence, it may happen that a knowledge engineer identifies at some later point of time that her/his understanding of a certain term was not yet right, that the term was misinterpreted, that a new term requires some disambiguation of previously modelled concepts or that some new super concept needs to be introduced for the new term or a previously modelled concept.

In subsequent modelling phases the modeller thus should also check and validate the previously modelled concepts and correct them if needed. For this task, the previously categorized terms function as entry points for the validation. While terms marked to be modelled (+) are natural entry points for validating and correcting the concepts of the terms environment, i.e. synonyms, super, sub- and other directly related concepts, terms marked as unclear (?) are natural re-entry points for the resumption of research and reviewing their meaning.

Although such a validation of the currently available ontology could also be done by some independent second person (following the four-eyes-principle), this usually introduces additional communication overhead, since modeller and validator need to identify and agree on a common understanding.

### 2.3.3  Broadening Phase

Sometimes it may happen that a modelled ontology needs to be adopted for a different application domain or that the objective of a semantic application using the ontology gets extended. In these situations, new categorical concepts may become important and an entire category of new terms need to be modelled, e.g. if an ontology for medical diagnosis should be recycled in an application for medical treatments, or if a thesaurus supporting job search should be adopted for a search of continued education offers, as was the case with Ontonym's thesaurus.

Of course, in these situations the ontology needs to be extended by corresponding concepts. Since concepts of these new concept categories are probably not yet present in the ontology, an entire concept branch needs to be opened for them and, if needed, the new concepts need to be related to other previously modelled concepts.

### 2.3.4  Comparison with Other Modelling Methodologies

At the time we started with the modelling of the HR thesaurus, a number of modelling methodologies were already worked-out and described, like Methontology, On-To-Knowledge, HCOME, DILIGENT, RapidOWL and COLM [5]. These methodologies were derived from an academic viewpoint, often oriented along well-known software development methodologies. Most of these ontology modelling methodologies were developed and used in academic settings, with only a few of them tested in some corporate context.

While all of them describe process models for modelling ontologies, in general they did not answer the questions: "where to get the technical terms about the concepts to model" and "how to effectively proceed in the modelling process". Hence, our approach can be considered a pragmatic continuation or extension of these methodologies.

Some of them are designed for distributed collaborative modelling by a number of persons (with different skills). Although this is in general desirable in order to capture a shared understanding of the terms needed, for the first phase of ontology modelling this is usually too expensive since the ontology and application based on it need to first prove their utility. Hence, in the first phase a lean approach like ours is preferable, followed later by a collaborative and if needed distributed modelling methodology.

Our pragmatical modelling approach has a close relationship to COLM because one of the authors of [5], Ralf Heese, one of our partners at Ontonym, was also involved in the early stages of the modelling and the development of our approach.

## 2.4   From Thesaurus to Ontology

From a historical viewpoint thesauri and ontologies appear to be incompatible methods for knowledge representation.

Thesauri were initially developed in the area of library and information science in order to index textual documents within a subject area with terms of controlled vocabulary by humans. They distinguish between broader terms and narrower terms, synonyms, antonyms, related terms, preferred and hidden names only, without defining these terms formally.

Ontologies were developed in the field of computer science, artificial intelligence and knowledge representation in order to logically formalize descriptions of the world for reasoning purposes. An extension of the well-known definition [6] defines an ontology as "a formal, explicit specification of a shared conceptualization" [7]. This definition implies that an ontology is based on some knowledge representation formalism. Research in the field of knowledge representation found that such formalisms require clear, formal semantics in order to draw valid conclusions about the world and to derive precise statements about the reasoning complexity of different language subsets and their inferential capabilities [8].

Although both kinds of formalisms appear to be incompatible, disciplined modelling of a thesaurus allows proceeding from a purely human-centred approach of vocabulary structuring to a formally sound approach of knowledge representation. This requires that every term modelled in a thesaurus should not only be considered to represent some language construct used by humans, but also should be considered as representing a class (or set) of objects.

### 2.4.1   Our Approach

Our overall objective was the development of a domain ontology for HR. However, it was clear from the beginning that the development of a neutral, complete, logically consistent ontology capturing all different views on the domain, which is applicable for a broad range of –yet unknown– applications, would require extreme efforts and would be very expensive. Without any justification, no one risks such an investment. However, we had in mind that someday we could come across a customer problem where such an ontology would be needed.

It was also clear (from first author's experiences at T-Systems) that a semantic search based on a lightweight knowledge representation could support users, if it accounts for their language use, "understands" synonyms and uses more specific and related terms to augment the search results. Therefore, instead of shooting into the dark with the development of a full-fledged ontology, we started with the development of a HR thesaurus. However, in order to allow for a future transition to an HR ontology, we developed the thesaurus under the following framework conditions:

- The representation formalisms should allow for the modelling of ontologies.
- The representation formalism should be standardized.
- The development should be supported by freely available tools.
- Meta-concepts of thesauri should be formulated in terms of the ontology formalisms used.

At the time we started the modelling, SKOS [9] was not yet finalized, so we decided to use OWL as the representation formalism [10] and Protégé [11] as the modelling tool.

### 2.4.2 Modelling Guidelines

In order to allow for a later transition from the thesaurus to an ontology we established the following modelling guidelines for our semantic search:

1. Noun terms are the first class citizens of the thesaurus.
2. Any thesaurus term should denote a set of objects and thus should be represented as concept, i.e as OWL class.
3. The concept subsumption of the representation formalism should be used to represent the broader/narrower relation of the thesaurus. Hence, other relations often represented as broader/narrower relations (like meronomies) need to be represented differently (see 10. below). The concept subsumption forms a term hierarchy as directed acyclic graph.
4. The concepts of the term hierarchy should represent the preferred names of a term, in order to establish a controlled vocabulary. These concepts are called canonical concepts.
5. Preferred names of a term are represented as labels of the canonical concepts (non-preferred names are either represented as synonyms (see 6.), hidden (see 9.) or abbreviations (see 12.))
6. Synonym relations represent equivalences between terms. Hence, synonyms are represented as equivalent concept using OWL's equivalent class of a canonical concept. These equivalent concepts are marked as synonyms (either by a boolean data property or subsumption under a concept "Synonym" of an "application ontology") in order to distinguish them from the canonical concepts.
7. Other synonym names of the term are represented as labels of equivalent concepts.
8. A language designator attached to a label is used to represent translations of the term name in the designated language. Labels without a language designator represent terms which are common in all languages.
9. Hidden names are represented as synonym concepts. The corresponding concept gets marked as "invisible" (either by a boolean data property or subsumption under a concept "Hidden" of an "application ontology"). Hidden names are just used for the identification and mapping of uncommon terms and common misspellings to the controlled vocabulary represented by the canonical concepts.

10. Related terms are represented by a general directed object property (called "related_to") relating the corresponding concept to some other concept. A general symmetrical object property (called "sym_related") is used as a convenient abbreviation to automatically establish two directed "related_to" relations between two concepts.
11. The top-most concepts of categorical concepts are marked as "categorical" (either by a boolean data property or subsumption under a concept "Categorical" of an "application ontology").
12. Abbreviations are represented as synonym concepts. The corresponding concept gets subsumed under a concept "Abbrev" of an "application ontology". Analogous to the hidden names they are used to identify and map abbreviations to the canonical concepts.

### 2.4.3  Modelling Patterns

Besides the usual super and sub concept relations, which are used to represent a controlled vocabulary in the form a thesaurus' broader/narrower term relations, the guidelines were extended by some modelling patterns we found useful for simplifying some processing tasks in the context of semantic search.

**Synonyms are represented as equivalent concepts of a designated canonical concept.** This allows on one hand to compile them into a hash table for mapping synonyms into the controlled vocabulary, which can be used for fast lookups during document analysis. On the other hand, it would allow using reasoning techniques of description logics to check the logical consistency of the ontology.

**Relations are represented by a general object property "related_to" in order to establish directed relations between concepts**. This object property can be used to establish relations that do not represent concept subsumptions, such as "part of" meronomies, "see also" relations or other domain-dependent relations. It is used to represent relations between concepts through value restrictions. These relations are especially interesting if a search application uses them in "reverse order". E.g. if we introduce an association that nurses are usually related to some hospital, i.e. 'nurse related_to some hospital' (in Manchester Notation of Description Logic [12]) then it is perfectly reasonable for a job search to return results containing the term 'nurse' if the user searches for 'hospital'. However, returning results which contain the term 'hospital' is probably inadequate for a search of 'nurse', since the latter is in a certain sense more specific. In that sense, the object properties are used in reverse order.

**Representation of ambiguous terms.** We identified two forms of term ambiguity patterns during the modelling: either two different concepts have a common subconcept or one term is used to refer to two different concepts. The former usually occurs if two different terms are used to describe two different perspectives of the same term, e.g. an 'offshore wind turbine' is a 'wind turbine' as well an 'offshore plant'. If these more general terms are searched, this kind of ambiguity usually causes no problems. The latter kind of ambiguity occurs if one term (e.g. an abbreviation) is used to denote two different terms, e.g. the abbreviation 'PR' denotes on one hand 'public relation' and on the other hand 'progressive relaxation' a synonym for 'progressive muscle relaxation' a therapeutical relaxation technique, or 'PDMS' which stands for 'patient data management system' as well as for 'product data management system'. These latter ambiguities can be resolved by introduction of an "application level" concept for abbreviations, which holds these ambiguous abbreviations and allows to search for both interpretations if the abbreviation is sought.

### 2.4.4   Refining the Thesaurus to an Ontology

Following the above guidelines, most of the conceptual components of a thesaurus are already mapped into the more expressible representation formalism of an ontology.

For the transition from a thesaurus to an ontology these definitions could of course be further extended by the introduction of range and number restrictions, by negation, disjointness or completeness statements. Beside these refinements which concern the definition of concepts, the relation "related_to" can be further refined by specialized subrelations, which of course still need to be defined, to substitute them in definitions with more precise meanings.

Since the "related_to" relation was used already to establish general relations between concepts by specifying the relation as a primitive concept definition of the form `related_to some TargetConcept` (Manchester Notation), it can be used to identify easily the concepts which need to be described by more precise relations.

Technically, the refinement of the thesaurus to an ontology thus is quite easy, although it may require still larger efforts.

---

## 2.5   Experiences

From 2008 until 2013 the HR Thesaurus was under active development. Since then it transitioned slowly from development into maintenance mode. During this time and over its usage for the realisation of several different semantic search applications, we have had some practical experiences, which we find worthy of sharing.

### 2.5.1   Modelling Effort

During the entire development and maintenance time of the thesaurus we continuously recorded the modelling effort needed by measuring the number of terms modelled per hour. Since the entire modelling was based on keyword lists as described in Sect. 2.3.1.2, it was easy to mark modelled terms, to track the hours spent for modelling, to count the terms modelled and summarize the total number of hours. It should be noted that during the modelling of one term, relations between terms and additional terms were modelled to define the term properly. The recorded time includes the inspection, validation and, if needed, the correction of previously modelled terms as well as time for occasional background research for the term's meaning.

Our experience has shown that without detailed and exhaustive background research for the meaning of terms it is possible to model 20–30 terms per hour. Background research here means, according to the right side of Fig. 2.2, to occasionally pose a Google search query if the meaning of the term is unknown or unclear, to inspect the first result page returned by Google and if needed some web page referenced from this result page for finding a common definition of the sought term.

In a strict sense, this empirically found value is valid only for the application domain of the HR thesaurus. However, we think that it does not depend on the domain itself, but instead on the task and application, i.e. to support the search process by accounting for the language use of users, for which the thesaurus is built.

Interestingly, the number of terms modelled per hour remained quite stable over the entire development period. However, we also experienced that the number of terms to model decreased over time and that more time was needed to validate and correct previously modelled concepts.

At first sight, 20–30 terms per hour appears quite small. Modelling 4,000 terms, i.e. some moderate thesaurus, would thus require 20 days of working time. Modelling the entire HR thesaurus took us in total 3 months of working time, distributed over 6 years with certain phases of extensive extensions.

By combining this empirically measured effort with the internal costs of your company, you get an estimate of how expensive the modelling of your terminology could be. But don't look at these costs in isolation. Instead, relate them to the cost of manually building up a database of equal size or to developing a program with an equal number of lines of code and it becomes apparent that modelling is not more expensive than conventional database or software engineering.

## 2.5.2   Meaning-Carrying Terms

One thing we found interesting during the text mining and modelling, was that for a defined application domain certain nouns and noun phrases carry little information and are thus not worthwhile modelling.

As explained previously, in every application domain important concept categories can be identified. Terms falling into these categories are prime candidates for the modelling in search applications since users will search for them. We call these terms *meaning-carrying terms*, since they carry the most information for document searches. These terms are often nouns and noun phrases.

Nouns and noun phrases can be identified quite easily with computer linguistic text engineering methods and tools (see e.g. Sect. 5.2). However, the derived keyword lists often consist of a mixture of such meaning-carrying terms and other noun phrases carrying less information for a particular application, but occurring frequently in the analysed documents. For example, in Sect. 5.4.1, while most of the terms which co-occurred with the term "compliance" are meaning-carrying, the terms "Groß", "Fragen", "Rahmen", "Jahr" or "Dauer" are of a very general nature and are not tied to the particular domain.

Other examples of such general terms which carry little information in an application domain include "object", "start", "end", "middle", "property", "process", "thing", "class". From the viewpoint of knowledge representation, these are general terms or concepts often defined in so called "upper ontologies" or "top-level ontologies" [13] from the viewpoint of applications such as semantic search, they carry too little information to be of interest for users.

In order to simplify the modelling process, it would of course be very useful if we could identify only the real meaning-carrying terms and to ignore such general terms such as stopwords. However, the only approach we are currently aware of for identifying meaning-carrying terms is by deciding whether their relative frequencies differ significantly from their frequency in some general corpus. But this leaves an open question: "which comparison domain/corpus to use?"

### 2.5.3   Spelling Tolerance

In different search applications ranging from intranet search engines, job portals for engineers, search engines for jobs and continued education for workers and a search application for children between the ages 6 and 15 for a TV broadcaster, we found that the users made a lot of typing and spelling errors. While simple errors like inclusion, omission, exchange and permutation of characters are quite easy to identify and in simple cases can be automatically corrected, we needed to augment our search function through specific solutions to cope with spelling errors for German compound terms, foreign words and names.

In German it is quite easy to create new expressions through compound terms consisting of a number of base terms. However, the rules for connecting these base terms are not simple and users tend to make mistakes, either by connecting terms which should be separated by a hyphen or by hyphenating terms which should be written together with a hyphen or space. For coping with all combinations of these errors, we developed an approach for generating all possible error combinations as some kind of hidden term in order to map those erroneous spelling variants to the correct controlled vocabulary term.

Additionally, we found in a search application for a German TV broadcaster that not only children but also adults have problems with the correct spelling of foreign names. For this application we additionally used a phonetic encoding for mapping different phonetic spellings into the controlled vocabulary.

### 2.5.4   Courage for Imperfection

Our incremental modelling approach is quite specialized and of course not applicable for every application. The approach itself is based on the requirement that the thesaurus is made operational as soon as possible in order to validate the model it represents, to acquire new terms from users during its operation and in order to avoid high modelling costs at the beginning, instead distributing them over the lifetime of the model. This approach thus requires a good amount of courage on the side of the search engine operator to stand the temporal imperfection of the model and to react if users complain.

Of course, in the beginning complaints and queries such as "why some particular piece of information could not be found with a particular search query" are unavoidable and require short reaction times to correct these faults. Most of these queries come from document providers validating that their documents can be found. However, we found that the number of such request remains quite low and diminishes over time as the model evolves.

There have been no reports from our clients of complaints from users that sought information could not be found. This probably depends on the imperceptibility of which information is indexed by the search engine.

## 2.6 Recommendations

Often we heard from practitioners in the field of semantic technologies, that the development of an ontology is too complicated, too expensive, requires too much effort, etc., leading them to switch from the use of handcrafted ontologies to available linked open data. However, for a large number of applications the available linked open data sources do not cover the terminology of the particular domain completely and what is more problematic: they do not cover the language use of users.

Although modelling of an ontology is often done by the developers of a system, this is obviously not their business. Hence, it seems reasonable to us that they disregard the modelling of thesauri or ontologies as requiring too much effort and instead switch to data sources which can be processed automatically. However, there are library and information scientist whose business it is to do terminological work, which are prepared and educated to carry out this kind of intellectual work.

Often the development of a correct, complete or consistent ontology is not needed for a search application. If the ontology covers 80 % of the most frequently used terms of a domain, a large number of users can already benefit from the effort of modelling the application domain's terminology.

Additionally, if a pragmatic development method is chosen and the ontology is put into operation as early as possible, the data derived from its operation can be used for widening the acquisition bottleneck of important and relevant terms. As our experience has shown, the time required for modelling does not imply unaffordable costs, it just requires patience and time to collect the terms.

Therefore, besides letting experts do the terminological work and adopting an incremental framework which supplies the needed terms, our main recommendation is to start with a simple model, show its usefulness, extend it later and do not try to achieve perfection from the beginning. Instead, try to establish an error-tolerant environment for the extension of the model.

## References

1. Hoppe T (2013) Semantische Filterung – Ein Werkzeug zur Steigerung der Effizienz im Wissensmanagement. Open Journal of Knowledge Management VII/2013, online under: Community of Knowledge.http://www.community-of-knowledge.de/beitrag/semantische-filterung-ein-werkzeug-zur-steigerung-der-effizienz-im-wissensmanagement/. Last access 2 Nov 2017
2. Robinson PN, Schulz MH, Bauer S, Köhler S. Methods for searching with semantic similarity scores in one or more ontologies. US Patent US 2011/0040766 A1, filed 13 Aug 2009, published 17 Feb 2011
3. Schulz MH, Bauer S, Köhler S, Robinson PN (2011) Exact score distribution computation for ontological similarity searches. BMC Bioinformatics 12:441. https://doi.org/10.1186/1471-2105-12-441. Last access 2 Nov 2017
4. Hoppe T (2015) Modellierung des Sprachraums von Unternehmen. In: Ege B, Humm B, Reibold A (eds) Corporate Semantic Web – Wie semantische Anwendungen in Unternehmen Nutzen stiften. Springer, Berlin
5. Luczak-Rösch M, Heese R (2009) Managing ontology lifecycles in corporate settings. In: Schaffert S et al (eds) Networked knowledge – networked media, SCI 221. Springer, Heidelberg, pp 235–248
6. Gruber TR (1993) A translation approach to portable ontologies. Knowledge Acquisition 5(2):199–220. Academic Press
7. Studer R, Benjamins VR, Fensel D (1998) Knowledge engineering: principles and methods. Data and Knowledge Engineering 25:161–197
8. Baader F, Calvanese D, McGuinness DL, Nardi D, Patel-Schneider PF (2003) The description logic handbook. Cambridge University Press, Cambridge
9. W3C. SKOS Simple Knowledge Organization System. https://www.w3.org/2004/02/skos/. Last access 2 Nov 2017
10. W3C. OWL 2 Web Ontology Language document overview, 2nd edn. https://www.w3.org/TR/owl2-overview/. Last access 2 Nov 2017
11. Protégé. A free, open-source ontology editor and framework for building intelligent systems. https://protege.stanford.edu/. Last access 2 Nov 2017
12. Horridge M, Drummond N, Goodwin J, Rector A, Stevens R, Wang HH. The Manchester OWL syntax. http://ceur-ws.org/Vol-216/submission_9.pdf. Last access 2 Nov 2017
13. Herre H, Uciteli A (2015) Ontologien für klinische Studien. In: Ege B, Humm B, Reibold A (eds) Corporate Semantic Web – Wie semantische Anwendungen in Unternehmen Nutzen stiften. Springer, Berlin

# Compliance Using Metadata

**3**

## Rigo Wenning and Sabrina Kirrane

**Key Statements**

1. Get conscious about the workflows and create a register of processing activities that are affected by compliance requirements (e.g. Privacy requirements).
2. Model the policy constraints from the legal and corporate environment into Linked Data to create policy metadata.
3. Attach the relevant policy metadata to the data collected, thus creating a semantic data lake via Linked Data relations or Linked Data annotations
4. Query data and relevant metadata at the same time to only process data that has the right policy properties.
5. Write the fact of processing back into the semantic data lake and secure it with appropriate measures (e.g. Blockchain).

R. Wenning (✉)
European Research Consortium for Informatics and Mathematics (GEIE ERCIM),
Sophia Antipolis, France
e-mail: rigo@w3.org

S. Kirrane
Vienna University of Economics and Business, Wien, Germany
e-mail: sabrina.kirrane@wu.ac.at

© Springer-Verlag GmbH Germany, part of Springer Nature 2018
T. Hoppe et al. (eds.), *Semantic Applications*,
https://doi.org/10.1007/978-3-662-55433-3_3

## 3.1    The Increased Need for Tools for Compliance

The digitisation of all aspects of our life results in systems becoming ever more complex. As they get more complex, humans have more and more difficulties in trying to understand what these system do. As billions of people live their lives online, they leave traces. Others have started to measure our environment in all kinds of ways. The massive amount of sensors now produces massive amounts of data. Moreover, because our society communicates in many new ways online, it creates new complex social models. The advent of open source software can be taken as an example. The open source ecosystem would not be possible without the Internet and the Web that allows complex governance structures to be built online [1]. Social networking, browsing habits and other interactions online are recorded. This leads to the creation of massive amounts of data. Data collection online and offline corresponds increasingly to the big data characteristics of velocity, variety and volume.

It is now tempting for certain actors to exploit the intransparency of those complex systems. This is done by harvesting and monetising data in opaque ways, or by just benefitting from protocol chatter. The internet as a whole has basic vulnerabilities in this respect. The most threatening example is certainly the pervasive monitoring of all internet traffic by the NSA and GCHQ [2]. Additionally, the private sector is monitoring behaviour on the web, also known as "tracking". A short term benefit is offered to a targeted individual and people do not realise the long term danger of the profile created. Various techniques are used to build profiles of people and sell them to the highest bidder. Entire platforms and toolchains are created and made available for free to be able to monitor what people do on a system. Some call this the surveillance economy [3] as the prices for ads targeted to a profiled person generate much higher revenue than normal banner ads.

Malicious behavior in complex systems is not limited to eavesdropping on communications. The recent scandals on IT manipulations revealed that by manipulating one end of a complex system, there can be huge benefits on another end of such a system. An example of this is the Libor scandal [4] in the financial industry, where an index used for the calculation of interests was gamed to obtain certain results. The software in cars also had hidden functions that detect when the car is in a test cycle inside a lab and changes the engine's characteristics to comply with requirements said to be unachievable.[1]

The combination of complex and intransparent systems with manipulations is undermining the trust people have in the correct functioning of those systems. This is especially true if sensationalist media with a hunger for audience and attention widely reports and exploits those topics. Verification by users is difficult.

As a consequence, people are more reluctant to use those complex systems. If they have an opportunity to avoid using the systems they will do so due to the lack of trust and confidence. This creates economic inefficiencies and hinders further progress of society by even more complex systems. The pace of innovation is seriously endangered if people start to mistrust the IT systems they use.

---

[1] Also known as the Volkswagen emissions scandal, but many vendors are implicated.

Consequently, governments all around the world create new regulations and demand compliance with those regulations. The aforementioned privacy abuses and tracking excesses have accelerated the reform of the European data protection law that led to the General Data Protection Regulation (GDPR) [5]. The scandals in the financial industry resulted in additional rules for reporting. However, lawmakers responsible for the regulations are often positivistic and underestimate the difficulties in implementing the regulation by transforming rules into code and organisational policies. Often, the compromise is to implement what is implementable and try to not get caught with the rest. The approach suggested here puts forward a different way. It suggests to use more technology: "*Social rules*" are translated into machine understandable metadata and can then steer and guide the behaviour of our complex systems using such metadata. "*Social rules*" in this sense include laws, usages but also user facing promises like privacy policies.

In order to demonstrate the compliance with regulations, especially in data protection and security, laws and implementation provisions very often recommend certification. The traditional way of doing certification is to engage some expensive consultant or auditor who examines the IT system and asserts that the system does what it says it does. The result is often an icon displayed on a website. This is very expensive and does not scale well. Additionally, those certification systems have a number of disadvantages. A slight change to the system can render a certification void.[2] Security certifications can even be harmful according to a study by KU Leuven [6]. The privacy seals carry a dilemma that a service showing the seal also pays the seal provider. The seal provider has little interest in going against their customer. Not only has the manual certification disadvantages, but it is also inflexible. In order to address the trust issue, the SPECIAL project[3] develops a standardised transparent Web-based infrastructure that makes data sharing possible without destroying user confidence. This will benefit the overall economic impact and growth of the data value chain.

With the metadata approach, a flexible system is installed that can cope with policy changes and audits. Creating such a system requires some scrutiny over business processes, data collection, purposes and retention times. Implementing the approach thus serves the goals of renovating and optimising the business processes at the same time.

## 3.2    Making the Data Lake Usable

Since the Primelife project[4] we know about the issue with data reuse in Europe. Typically, data is collected at some point in time under certain conditions. Then this data is written into some data warehouse. Later, there is no knowledge anymore under which conditions the data was acquired and recorded into the data lake.

---

[2] Except where the certification is meaningless or very imprecise.

[3] www.specialprivacy.eu.

[4] http://primelife.ercim.eu/.

All those warehouses create a data lake full of valuable data. However, the data lake now contains a lot of muddy water; as there is no information about the purposes for which the data was collected, nor any information about the rights attached to the data. A potential new use of that valuable data faces the obstacle of legal uncertainty. This legal uncertainty creates a sufficient commercial and liability risk to deter commercial actors from realising the potential of the data lake. Regarding privacy, we can see that those services not following data protection rules can monetise their data and grow, but they will erode trust and usage of data driven IT services. The eroded trust will decrease the use of those services and generate less data, or it creates lower quality data because people lie about things. In our digitised world, networked services generate lots of data. Using this data is the most promising source for higher productivity and wealth. To avoid the erosion of trust and the decline of networked services, systems have to demonstrate how they follow the democratically established rules. But in the era of big data, the complex systems are too complex for pure human cognition. This means technology needs to help achieve demonstrable compliance that is understandable by humans.

Data protection often calls for data minimisation, meaning the collection and storage of less information. However, the suggestion here is to add even more information. This additional information is metadata containing policy and usage information that allows a system to keep the promises made to the data subject. It allows the data lake to be both usable and compliant. By having this information available, the risk and liability can be assessed. At the same time, it allows for a better automation of compliance processes required by law because it adds a layer of transparency and logging.

## 3.3    Concept and Architecture of a Policy Aware System

### 3.3.1    Data Acquisition

The suggestion is to make the "*social rules*" available to the system machine readable as far as possible[5] We refer to this as "*policy information*" in the following. This is best done using Linked Data (See Chap. 4). The policy information completes the "*environmental information*" available at data acquisition time. Environmental information here means all protocol headers and other information available to the IT system in question at data acquisition time. This information is normally spread over a variety of log files, but may also include other information sources like policy files and DNT headers e.g. Now the system has not only collected the actual personal data, but also metadata about the collected personal data. This could also be metadata about financial data or other data relevant for compliance with given policy information. The idea behind this extensive data acquisition is that, at collection time, we normally know perfectly well under which conditions data is acquired and, e.g., for how long it can be held. Today, this information is lost and forgotten once data is transferred to some data warehouse. However, the system suggested

---

[5] See [7] and [8] were projects trying to implement some part of the idea.

here will remember the constraints at collection time. Those constraints are stored as metadata and connected to the data collected. For various cases, there are existing ontologies to allow the conversion of environmental and contextual information into machine readable metadata. For other cases such ontologies or sometimes even taxonomies can be created for the specific requirements of a given business process.

### 3.3.2   Connecting Data and Metadata

Now we have the payload data[6] that is subject to the intended processing, and we have metadata that tells us more about the data items processed, how long we can retain them, the purpose of the processing and other promises made to the user at acquisition time. Acquiring a full policy into an SQL database with every data item collected would be overkill. Instead, we need to link the metadata to the actual payload data, e.g. personal information like location data and mobile number.

For the system to work, the acquired payload data has to be transformed into Linked Data, a process commonly known as "semantic lifting". Semantic lifting aims at adding "meaning" or extra meta (semantics) to existing structured/semistructured data following Linked Data principles and standard Semantic Web technologies [9]. A key feature of Linked Data is that it uses IRIs [10][7] to identify data uniquely. Once the payload data is identified uniquely, the metadata can point to it.

For example, let us imagine the acquisition of a mobile number for some service that has to be erased after 3 weeks. The IRI given to the mobile number at the semantic lifting could be http://wenning.org/ns/mtel/12345678[8] and the IRI given to the rule for the 3 weeks data retention could be http://www.w3.org/2002/01/p3prdfv1#retention, which equals 1814400 seconds according to the P3P vocabulary. The triple then indicates the phone number, the retention attribute and the retention time. As these are globally unique identifiers, this even works across enterprise boundaries.

Once the semantic lifting is done and once all payload data recorded is given a IRI, the policy information pointing to the IRI of the payload data record can be seen as an annotation of that record (Fig. 3.1).

While the W3C Annotation Recommendation [11] is about adding remarks on web pages, we annotate data records, but as a web page is a resource with a IRI, the principle is the same. The Annotation data model [12] (Fig. 3.2) consequently states: "An annotation is considered to be a set of connected resources, typically including a body and target, and conveys that the body is related to the target. The exact nature of this relationship changes according to the intention of the annotation, but the body is most frequently somehow 'about' the target". This perspective results in a basic model with three parts, in both cases.

---

[6] Payload data means the actual data record, e.g. the name of a customer.

[7] IRI – Internationalized Resource Identifiers, the international version of URI according to RFC.

[8] The IRI for the mobile number is a purely theoretical example, the retention time is from the P3P 1.0 Specification.

**Fig. 3.1** Annotating data with metadata using RDF

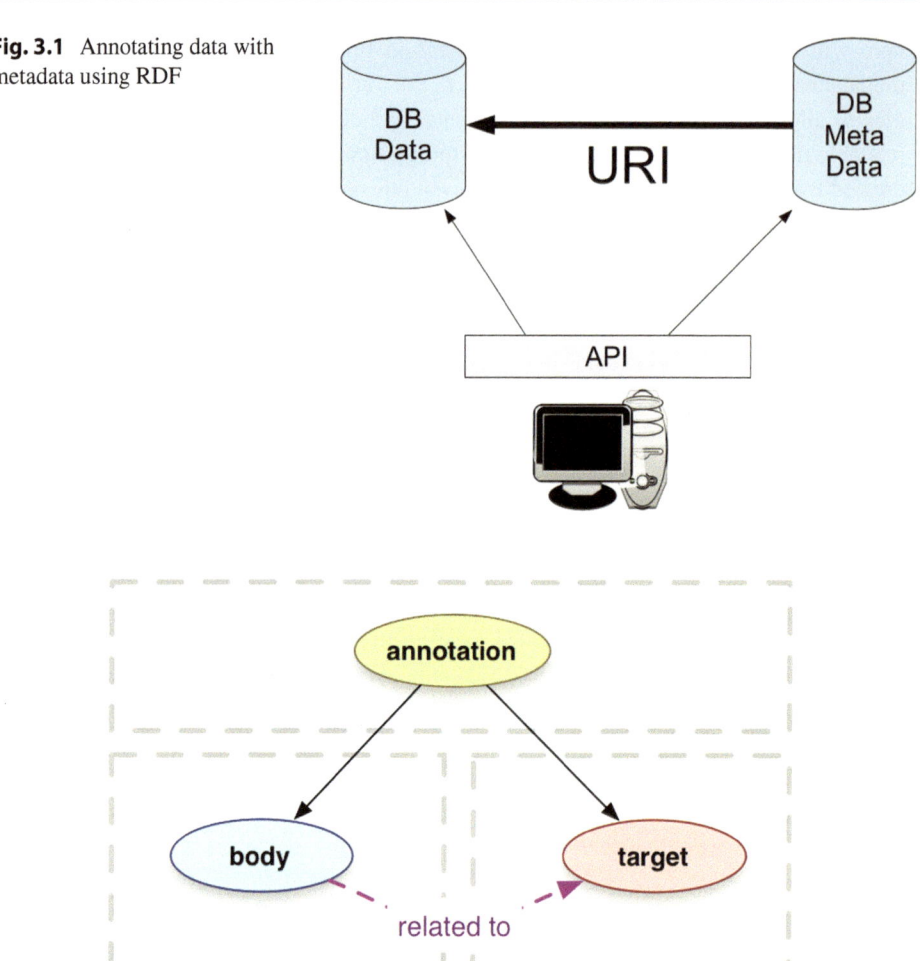

**Fig. 3.2** The W3C Annotation data model. See https://www.w3.org/TR/annotation-model/. (Copyright © 2017 W3C® (MIT, ERCIM, Keio, Beihang). W3C liability, trademark and document use rules apply.)

During data collection, the environmental information (metadata) and the data receives a IRI each, which allows for linking them together in a triple. This is a form of semantic packaging. Through this semantic packaging, algorithms of the system can react based on that metadata. This rather simple idea leads to a significant amount of social and technical challenges that are not as simple as the basic idea. In the following, the concept is exemplified by a system that tries to achieve compliance with Regulation (EU) (2016) 2016/679 [5]. The SPECIAL H2020 research project explores ways to implement such compliance and make big data privacy aware.

The system will also work to implement the data retention framework of DIN 66398 that goes beyond privacy and also takes commercial archiving duties into account. It is also perfectly possible to implement reporting and diligence rules from the financial sector regulations. Of course this can be all done using a SQL database with hardwired semantics, but this would remain inflexible and become an island that would not be capable of connecting to the complex world surrounding it. This also means that we need Linked Data in order to allow for data value chains that can be adapted without reprogramming the entire system.

### 3.3.3 Applying Constraints

As the system is using Linked Data, queries will be formed using SPARQL [13]. Once the system has data and metadata ingested, the compliance is a matter of applying the right query. It is now possible to make intelligent and policy-aware queries like: "find all data that can be further processed to provide a personalised service". It also allows, according to DIN 66398, to say: "list all data where retention time ends within the next week". In order to make such a query, of course, the system has to know about retention times. An important category of query will concern the constraint on data sharing by asking the system to return only data, for example, "that can be shared with business partner B1 for purpose P2".

### 3.3.4 Creating Policy Aware Data Value Chains

Division of labour creates higher efficiency and adds value. In the digitised economy, division of labour also means sharing data. We know that the "sharing economy" created a lot of enthusiasm. At the same time people are more and more concerned about data sharing because the meaning of sharing remains unclear. Is data shared for any use? A solution may be to share data with the policy or constraint-information attached. This is not a new idea. It was most probably Michael Waidner[9] who coined the term "sticky policies" for this concept.

As mentioned, the use of IRI's allows for the preservation of the bundle of policy information and the payload data, even in a collaboration scenario. For a typical data protection scenario, the PrimeLife language uses the terms "data subject", "data controller" and "downstream data controller" (Fig. 3.3).

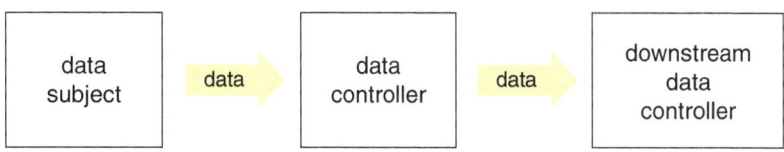

**Fig. 3.3** A data value chain from a privacy perspective

---

[9] IBM Zürich at the time, now Director of Fraunhofer SIT in Darmstadt.

It is then a matter of choice and of the business model attached to the intended data flow, whether the metadata with policies, constraints and obligations is:

1. Packaged together with the data records in various ways.
2. Delivered in two independent files.
3. Made available via some API by the data controller to the downstream data controller.

The downstream data controller will then have to apply the same constraints as the data controller. Securing this relation is a matter of contractual provisions proscribing the adherence to the constraints received or open to cryptographically secured systems. Such cryptographically secured systems can be similar or even close to the rights management and rights labelling systems we know already today.

### 3.3.5   An Automatic Audit for Compliance

If a certain personal data record is processed, this can be written as usual into a log file. Instead of using a log file, this fact can also be part of the metadata of the system. In this case the fact of collecting and processing that payload data is attached as an annotation to the payload data. The system can now reason over the meaning of terms like "purpose". With this in mind, a log of relevant information about processing, purpose of processing, disclosure and sharing can be held. Because it is not only a log file, but rather something that can be queried in sophisticated ways, the audit itself becomes a matter of a query into a certain stream within a business application.

It is now possible to ensure that the data controller actually did what they claim that they did. To ensure this, ensuring the security and the integrity of the log file is essential. This can be done by third parties, by some corporate rules or organisational measures. The modern way would be to use blockchain technology and write transactions and meaningful processing on that secured ledger. This is what the SPECIAL project's research is also focused on.

### 3.4     Finding and Formalising the Relevant Metadata

A policy-aware system is only as useful as the policy information recorded as metadata. The primary obstacle to the realisation of the presented vision is the lack of reusable and machine-readable context information of sufficient quality linked to the actual data being shared. Additionally, policies are often defined only generically on paper. Likewise the consent to use data is often only collected on paper, containing an entire wealth of information and thus not capable of being specific enough to decide on real use limitations or data handling directives. It is important to provide an infrastructure that allows human-readable policies to be tied to their machine-readable counterpart. This will preserve context information and usage limitations and transport them with the data.

It is a huge task to model and formalise policy information and to allow for the semantic lifting described above. The complexity and richness of the semantic lifting depends on the variety of metadata added to the payload data, and in how far taxonomies and ontologies are already available for that use case. Here there is a strong interest in standardisation as it will allow for a wide reuse of the vocabularies established by those specifications. This may include industry codes of conduct, but also policy snippets that may be combined in new ways.

For privacy use cases, some progress has been made by the creation of the PrimeLife[10] [8] language extending the XACML language [14] to allow for the use of credentials and interoperable role based access control. For security assertions and access control data, the semantics of SAML [15] can be used. The SPECIAL project has a focus on using ODRL [16] to express usage constraints and obligations attached to the collection of data. A good way to add metadata about quality of data is to use the W3C Provenance framework [17]. For financial services the work has not been done yet, but looking into the semantics of the eXtensible Business Reporting Language XBRL [18] may help. Chapter 2 shows ways to approach this pragmatically.

Once the taxonomy or ontology of a given policy is identified and modeled, a receiving entity or downstream controller can receive the schema and thus adapt their system quickly to the requirements for cooperation with the data controller. A large variety of data sources and types are expected. The more that digitisation advances in our lives, the more context information will be available to the system. All this information will feed into and further increase the data lake. It is therefore of utmost importance to keep in mind the variety management described in Chap. 4. So far, most information is used for profiling and marketing. The justified fear is that to know your customers means to be able to manipulate them to their detriment. The proposal here is to use more data to give users more control.

## 3.5    Context-aware Reaction as a Decisive Leap to Usability

A system that collects policy data at collection time knows about the constraints, the process under way, the purpose of collection and many more aspects. The current state of the art in compliance and information is marked by two extreme ends. On the one end, there are complex and lengthy information sheets provided to people. Many people have received pages of information from their bank that could later be used to excuse foreclosures. McDonald et al. [19] found that the average privacy policy has about 2500 words. The results of their online study of 749 Internet users, led the authors to conclude that people were not able to reliably understand privacy practices [20].

One of the reasons for this lack of understanding on the users behalf is an issue that already surfaced in privacy policies and also in a financial context. An all encompassing privacy policy or notification makes it necessary to cover an entire operation with all

---

[10] See footnote 4.

available branches and possibilities in one document. This leads to a legal document with a "one size fits all" mentality that is not geared towards the user, but entirely dedicated to the avoidance of liability. Lawyers are used to such long documents, but those documents are a disaster for usability. More and more voices declare the failure of privacy policies to generate trust while acknowledging that they are effective from a liability avoidance perspective.

The SPECIAL system allows for a radical change to this approach by allowing the building-up of an agreement between users and data controllers over time, step-by-step. This will serve user confidence and trust and, with the appropriate add ons, protect against liability. It will achieve this via the use of Linked Data, where a graph is created in which nodes are connected to each other. In a stateful system a given point has information about the nodes surrounding it. The system can use this information about the surrounding nodes to create a context-aware user experience. Instead of the entire policy, the system now knows which parts of the policy, constraints or obligations currently apply. The user and commercial partner interfaces do not need to display all information, but can rather concentrate on relaying relevant information with respect to the current interaction. This is helped via categorisation during the modeling phase of the policy at collection time. The categories can be reused to help the interface achieve a layered information interface.

Applied to the General Data Protection Regulation [5], the system shows the relevant and required contextual information. While a general policy is just an informational document, the SPECIAL system now allows for direct interaction. By implementing a feedback channel, an affirmative action of the data subject can be gathered from the interface, leading to an agreement the GDPR calls "consent". As the system is policy aware, it can store the consent back into the data lake and into a transparency log or ledger. The user may encounter such requests for consent in different situations. For every single situation, experience shows that the actual request is rather simple: "We will use your login credentials to display your profile identity to others in the forum for mutual help. This forum is not public". Over time, the user may be contacted to agree to some "research for trends within the online forum". Additionally it would be highly desirable to allow the data subject access to this information within a dashboard. There is even research on how to cumulatively add this policy information. Villata and Gandon [21] propose a mechanism to combine permissions from various licenses into an overall set of permissions. It is therefore possible to accumulate a variety of context dependent permissions into a known and machine readable set of constraints, permissions and obligations that will govern the relation either to the data subject or the downstream controller. Applied to financial or other contexts, the interactions will generate a cumulative set of machine-readable agreements that can be automatically implemented by the system itself with demonstrable compliance.

Being dynamic will enable policies that apply at different levels of granularity to be tied to different parts of the data with different levels of sensitivity. This facilitates removing the need to fix the purpose for data collection upfront, by allowing both monitoring and control by the data subjects: by using machines to help humans overcome their cognitive weaknesses in big data contexts.

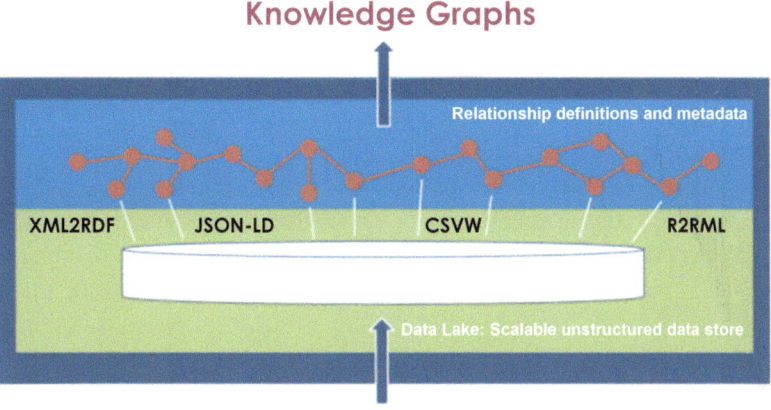

**Fig. 3.4** The semantic data lake. (Copyright: The Big Data Europe project https://www.big-data-europe.eu/semantics-2/ (accessed 20 Oct 2017))

## 3.6    Tooling for the Compliance System

The system described above needs good tooling. If data and metadata are recorded, this creates such a massive flow of data that big data technology is needed. Most big data tools today are not well suited for Linked Data, but after 3 years of development the Big Data Europe project [22] has created a platform capable of dealing with Linked Big Data.

BDE is first of all a normal big data platform using docker containers to virtualise data processing units and docker swarm to orchestrate those into a workflow. It created ready to use dockers for most of the Big Data toolchain from the Apache foundation. BDE calls this the Big Data Integrator (BDI)

On top of the BDI, tools for semantic operations have been created. Not all of them are production ready, but further development is advancing rapidly. BDE ended in 2017 but development of the tools continues. As a user of this technology, the SPECIAL project will also further the development of the semantic toolchain and help an already well established community. In the following, these semantic tools of the BDI are explained.

### 3.6.1    Tools for the Semantic Data Lake

Challenges in the suggested system are comparable to the Big Data challenges, namely volume, velocity, variety and veracity. Volume and velocity are largely solved by components such as HDFS, Spark and Flink [23]. However, in the BDE and SPECIAL use cases, variety is the biggest challenge[11] A lot of different data types and non-matching terms in different datasets are found. As discussed previously, the best way is to tackle the problem of variety is head on using Semantic Web technologies (Fig. 3.4).

---

[11] See Chap. 4.

### 3.6.2   Ontario or Transforming Ingestion?

BDE uses the "Ontology-based Architecture for Semantic Data Lakes" (Ontario) [24] (Fig. 3.5). Data is stored in whatever format it arrives in, but it can be queried and analysed as if it were stored as RDF. Ontario has the option to accept SPARQL queries that are then re-written and run over one or more datasets in whatever the relevant query language may be. The results are combined before being returned as a single result set. SPECIAL also has the option of transforming ingestion of relevant data and metadata. In this case, the raw payload data is semantified by unique identifiers to make it addressable by annotations.

### 3.6.3   SANSA Allows for Semantic Analysis

The SANSA stack (Fig. 3.5) [25] is a toolset that helps to streamline querying and reasoning over Linked Data. As we have seen previously, payload data and metadata annotations are stored in the system. Compliance is achieved through the filtering of data before the application of the intended processing. This means that the query/filter needs some level of sophistication to recognise the usable data sets. The SANSA Stack uses RDF data as its input and is able to perform analysis, e.g. querying, reasoning or applying machine learning over the Linked Data available in the platform.

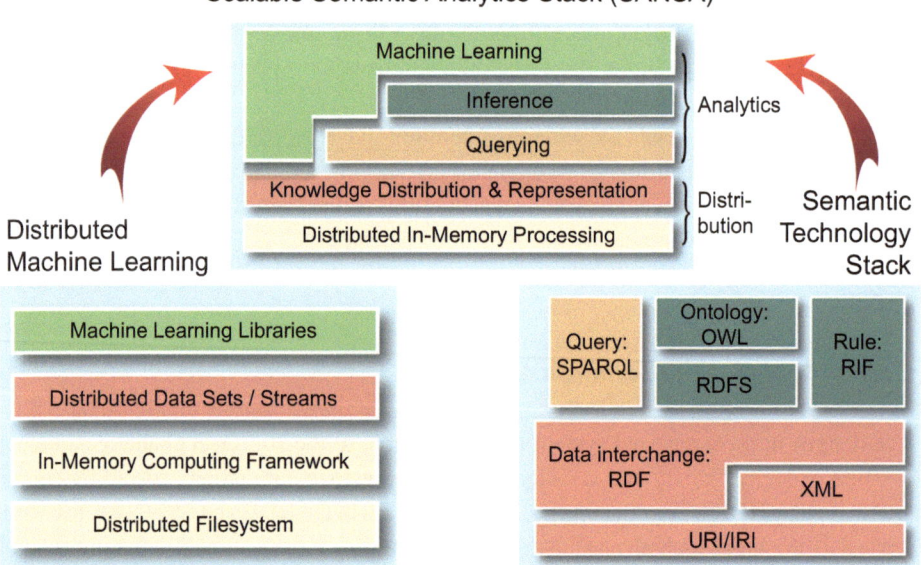

**Fig. 3.5** The SANSA stack. (Copyright: CC-BY Jens Lehmann http://sansa-stack.net (accessed 20 Oct 2017))

This allows exploration of not only the relation of payload data to metadata, but also the knowledge from the relations within the payload data or within the metadata. It helps construct the complex SPARQL queries needed to take account of permissions and constraints by providing algorithms that can be integrated into more complex and larger scale systems. Those will also typically need the parallelisation provided by BDE. It is still a challenge to parallelise SPARQL queries and reasoning. SANSA, although still under development, is well integrated into the BDE eco-system and provides docker-compose files as examples on github. This makes installation easy.

## 3.7    Recommendations

The advent of the GDPR will force companies to rethink their workflows. The time to think about adding a semantic layer for compliance is now. To do that:

1. Do the semantic lifting by giving IRIs to your payload data. The legacy systems can remain untouched as the IRI can point into it via some middleware.
2. Create the necessary taxonomies and ontologies according to Chap. 2 to allow for the appropriate semantics in the data annotations required for compliant data handling and checking.
3. Include respect for the metadata (annotations) provided into the contracts with business partners to insure respect for the data handling constraints.

## 3.8    Conclusion

The digitisation of our lives is progressing at high pace. The more aspects that turn digital, the more data we produce. The big data ecosystem depicts a situation where all the little streams from various ends form a big river of data: data that can be useful to fight diseases, but data that can also be used for manipulation. The possible options of dual use drive the call for more regulation. Data protection is only one field where the described system can be used for regulatory compliance. After the recent problems in the financial sector, new regulations about compliance and reporting were created. The SPECIAL system proposes the use of even more data to create a system of provable compliance. It also provides the basis for better integration of data subjects and users into the big data ecosystem by providing a dashboard and by organising a feedback channel. Additionally, it simplifies compliance management by private entities, and eases the task of compliance verification by data protection authorities. In essence, this brings trust to complex systems, thus enabling all big data stakeholders to benefit from data and data driven services.

# References

1. Raymond ES (1999) The cathedral and the bazaar: musings on Linux and open source by an accidental revolutionary. O'Reilly Media, Cambridge. ISBN 1-56592-724-9
2. A W3C/IAB (2014) Workshop on Strengthening the Internet Against Pervasive Monitoring (STRINT), London, 28 Feb–1 Mar. https://www.w3.org/2014/strint/. Accessed 20 Oct 2017
3. Lipartito K (2010) The economy of surveillance. MPRA paper, vol 21181, Mar. https://mpra.ub.uni-muenchen.de/21181/1/MPRA_paper_21181.pdf. Accessed 20 Oct 2017
4. https://en.wikipedia.org/wiki/Libor_scandal. Accessed 20 Oct 2017
5. Regulation (EU) (2016) 2016/679 of the European Parliament and of the Council of 27 April on the protection of natural persons with regard to the processing of personal data and on the free movement of such data, and repealing Directive 95/46/EC (General Data Protection Regulation), Official Journal of the European Union 59(L 119), May 2016, 1–88 ELI: http://data.europa.eu/eli/reg/2016/679/oj. Accessed 20 Oct 2017
6. Clubbing Seals (2014) Exploring the ecosystem of third-party security seals. In: Van Goethem T, Piessens F, Joosen W, Nikiforakis N (eds) Proceedings of the ACM SIGSAC conference on computer and communications security, Scottsdale. https://lirias.kuleuven.be/bitstream/123456789/471360/1/p918-vangoethem.pdf. Accessed 20 Oct 2017
7. Seneviratne O, Kagal L, Berners-Lee T (2009) Policy-aware content reuse on the web. In: ISWC 2009. http://dig.csail.mit.edu/2009/Papers/ISWC/policy-aware-reuse/paper.pdf. Accessed 20 Oct 2017
8. The PPL language, Primelife Deliverable D5.3.4 – Report on design and implementation. http://primelife.ercim.eu/images/stories/deliverables/d5.3.4-report_on_design_and_implementation-public.pdf. Accessed 20 Oct 2017
9. Tools for semantic lifting of multiformat budgetary data. Deliverable D2.1 from Fighting corruption with fiscal transparency. H2020 project number: 645833. http://openbudgets.eu/assets/deliverables/D2.1.pdf. Accessed 20 Oct 2017
10. RFC3987 Internationalized Resource Identifiers. https://tools.ietf.org/html/rfc3987
11. The W3C Web Annotation Working Group. https://www.w3.org/annotation/. Accessed 20 Oct 2017
12. Web Annotation Data Model, W3C Recommendation 23 February (2017) https://www.w3.org/TR/2017/REC-annotation-model-20170223/. Accessed 20 Oct 2017
13. SPARQL(2013) Query language for RDF, W3C Recommendation 21 March. http://www.w3.org/TR/2013/REC-sparql11-query-20130321/. Accessed 20 Oct 2017
14. See eXtensible Access Control Markup Language (XACML), currently version 3, with various specifications. https://www.oasis-open.org/committees/tc_home.php?wg_abbrev=xacml. Accessed 20 Oct 2017
15. Security Assertion Markup Language (SAML) v2.0 (with further info). https://wiki.oasis-open.org/security/FrontPage. Accessed 20 Oct 2017
16. ODRL Vocabulary & Expression, W3C working draft 23 February (2017) https://www.w3.org/TR/vocab-odrl/. Accessed 20 Oct 2017. See also the linked data profile https://www.w3.org/community/odrl/wiki/ODRL_Linked_Data_Profile. Accessed 20 Oct 2017 and the various notes linked from the WG page https://www.w3.org/2016/poe/wiki/Main_Page. Accessed 20 Oct 2017
17. An Overview of the PROV Family of Documents, W3C Working Group Note 30 April (2013) http://www.w3.org/TR/2013/NOTE-prov-overview-20130430/. Accessed 20 Oct 2017
18. XBRL 2.1. https://specifications.xbrl.org/work-product-index-group-base-spec-base-spec.html. Accessed 20 Oct 2017
19. McDonald AM, Cranor LF (2008) The cost of reading privacy policies, ISJLP 4, HeinOnline, 543. https://kb.osu.edu/dspace/bitstream/handle/1811/72839/ISJLP_V4N3_543.pdf. Accessed 20 Oct 2017

20. McDonald AM, Reeder RW, Kelley PG, Cranor LF (2009) A comparative study of online privacy policies and formats. In: Privacy enhancing technologies, vol 5672. Springer. http://dblp.uni-trier.de/db/conf/pet/pets2009.html#McDonaldRKC09. Accessed 20 Oct 2017
21. Villata S, Gandon F (2012) Licenses compatibility and composition in the web of data. In: Proceedings of the third international conference on consuming linked data, vol 905, pp 124–135. https://hal.inria.fr/hal-01171125/document. Accessed 20 Oct 2017
22. Big Data Europe. https://www.big-data-europe.eu. Accessed 20 Oct 2017
23. Components supported by the Big Data Europe platform. https://www.big-data-europe.eu/bdi-components/. Accessed 20 Oct 2017
24. Auer S et al (2017) The BigDataEurope platform – supporting the variety dimension of big data. In: Web engineering: 17th international conference, ICWE 2017, Rome, 5–8 June 2017, Proceedings, pp 41–59
25. SANSA – Scalable Semantic Analytics Stack, open source algorithms for distributed data processing for large-scale RDF knowledge graphs. http://sansa-stack.net/. Accessed 20 Oct 2017

# Variety Management for Big Data

4

Wolfgang Mayer, Georg Grossmann, Matt Selway, Jan Stanek,
and Markus Stumptner

**Key Statements**

1. Ontologies can aid discovery, navigation, exploration, and interpretation of heterogeneous data lakes.
2. Semantic metadata can help describe and manage variety in structure, provenance, visibility (access control) and (permitted) use.
3. Ontologies and comprehensive metadata catalogs can simplify interpretation, lift data quality, and simplify integration of multiple data sets.
4. Governance mechanisms for ontology evolution are required to sustain data quality.

## 4.1 Introduction

With big data applications reaching into all areas of corporate and individual activity, old challenges that were encountered in traditional application areas raise their head in new ways. Of the core challenges originally associated with big data, namely volume, velocity, and variety [1], variety remains the one that is least addressed by the standard analytics architectures.

W. Mayer (⊠) · G. Grossmann · M. Selway · J. Stanek · M. Stumptner
University of South Australia, Mawson Lakes, Australia
e-mail: wolfgang.mayer@unisa.edu.au; georg.grossmann@unisa.edu.au; matt.selway@unisa.edu.au; jan.stanek@unisa.edu.au; mst@cs.unisa.edu.au

© Springer-Verlag GmbH Germany, part of Springer Nature 2018                47
T. Hoppe et al. (eds.), *Semantic Applications*,
https://doi.org/10.1007/978-3-662-55433-3_4

According to the Big Data Execution Survey 2016, 69% of organizations consider handling variety of data as the prominent driver for the success of big data (while 25% named volume and 6% termed velocity as the major driver) [2]. Respondents agreed that bigger opportunities are found by integrating multiple data sources rather than by gathering bigger data sources. Challenges raised by volume and velocity can be addressed with sufficient processing power, increased networking capability, storage capacity, and streaming architectures. Hadoop has already pioneered the velocity problem through distributed computing. However, variety remains a significant challenge that cannot be addressed by better technology alone.

One of the salient features of big data platforms is that most are schema-less, that is, without a canonical description of what is actually contained in them. The absence of semantic information describing the content of the data stores poses difficulties for making effective use of data, in particular if data sets are contributed to and used by different user groups over prolonged time spans and processed in different ways. The value of metadata and semantic annotations is hence swiftly becoming a key consideration with respect to designing and maintaining data lakes. Due to diversity in its data contents, big data brings new challenges to the field of metadata management. Implementation of metadata management frameworks to support both data- and processing variety are key ingredients for providing greater visibility, data consistency, and better insight from data.

## 4.2    Big Data Variety

This section defines variety and analyzes types and sources of variety in big data. We introduce the concept of semantic metadata as a basis for describing and managing variety in the context of big data. Sufficiently powerful semantic metadata can capture the sources of variety in heterogeneous big data and record, once established, the mappings and transformations required to bring multiple data sources together in a joint usable pipeline. Management of these metadata must itself be closely aligned with that of the data, as their lifetime will extend as long as (or longer than) the actual data sources and results are used.

Issues related to data variety are of paramount importance today. Variety in stored data sets complicates comprehensive analysis, and variety in the underlying information systems and their interfaces impedes interoperability. We discuss variety from the following perspectives:

1. structural variation (by storage structure, data format, or semantic variation),
2. variations in granularity (either by aggregation or along the temporal axis),
3. heterogeneous sources of data,
4. degrees of quality and completeness, and
5. differences in data (pre-)processing.

We focus on the heterogeneity inherent in the data and leave aside policy and governance issues, such as access control and visibility of sensitive and confidential data elements. For discussion of policy related issues refer to Chap. 3. Strategies to address variety in data lakes are presented in the subsequent sections in this chapter.

### 4.2.1   Structural Variety

Data can exhibit structural variety in the form of varying data type, data format, and semantic variation.

Variation in data type is commonplace where data are held in multiple data sources. Different data types may be used to represent the same data, such as number representations with different bit-widths, strings of varying lengths, or textual and binary encoding of data values. These differences can often be overcome by transformations that convert from one representation to another when data are moved from one system to another. However, the conversion must be crafted and monitored carefully to ensure that data is transformed correctly and that no information is lost in the process.

Structured data has a fixed pre-defined format and thus is easily manageable. Semi-structured data does not have fixed structure, but does have some fixed properties which makes it relatively easier to analyze as compared to unstructured data. Unstructured data lacks structure and is the most common type of data today. Unstructured and semi-structured data is the fastest growing data type today and is estimated to comprise 80% of the total data [3]. Un-structured data is difficult to analyze in raw form. It needs to be prepared for analysis by various stages of pre-processing which may involve data cleaning, information extraction, categorization, labeling, and linking with other information.

More recently, graphs have emerged as a common data representation. The Resource Description Framework (RDF) and linked data are flexible approaches to representing data entities and their relations, and knowledge graphs derived from very large corpora of unstructured data facilitate analysis and knowledge discovery. Extracting information from all aforementioned types of data is important to enable deep analysis.

Semantic variations are perhaps among the most difficult to address, as subtle differences in meaning of the data may be challenging to detect. It is commonplace that data sets are mere snapshots of data sets that have been extracted from various sources as-is, with little or no explanation of the content and meaning of individual parts of the data set. The context and provenance of data are often lost in the process, which renders it difficult to make effective use of the data set for analytic purposes. If the process, purpose, and agent of data collection are unknown, potential biases in the data and subtle differences in meaning may be impossible to detect. Here, comprehensive metadata and provenance will be most useful.

Semi-structured, unstructured, and graph-structured data require no schema definition. It is hence tempting to focus on the provision of data in a data lake and neglect the associated metadata, instead relying on informal comments and the users' implicit knowledge to make effective use of data. Unfortunately, this approach breaks down quickly when the users contributing the data are different from those analyzing it. Even if each data set is described by an ontology, determining matching elements in different ontologies remains a difficult unsolved problem. Matching based on structural aspects and lexical representation is often imprecise, and explicit assertions of equivalences between concepts that have been specified in one context may not carry over to another context. Moreover, different conceptualizations and varying granularity of the concepts among ontologies in the same domain pose further challenges to matching and reconciling ontologies and their associated data sets. Changing ontologies and governance of their evolution also require further study.

### 4.2.2   Granularity Variety

Data can be viewed at multiple levels of aggregation. Data warehouses are organized such that raw data is aggregated along different dimension and granularity levels to enable analysis to efficiently pose queries at multiple levels of abstraction. The aggregation of data can occur at the value scale, where a precise value is replaced with more general representations; at the temporal scale, where elements at different points in time are combined into a single element representing a time span; spatially, where elements with a spatial region are aggregated into a single element representing the region; and along various domain-specific relationships. It is however important to know how data was collected and how it was aggregated, as different aggregation methods may support different conclusions.

### 4.2.3   Source Variety

Disparate data sources are among the major contributors to the variety aspect of big data. Data is being continuously generated by users, machines, processes, and devices and stored for further analysis. Deriving effective conclusions often requires combining data from multiple sources, integrating and linking data, and applying analytic methods to the combined data. Large-scale data analysis platforms often consist of an entire ecosystem of systems that store, process, and analyse data.

Heterogeneous data sources constitute an extra layer of complexity in big data ecosystems. The key challenge is how to integrate these sources to extract correlated meaningful information from them. With the advent of non-relational storage technologies, such as NoSQL and NewSQL data-stores, the ability to store semi-structured and unstructured data has increased. However, the need to overcome structural variety as described in the previous paragraphs is pressing as never before. In contrast to the advances in machine learning and automation techniques, the practices of software engineering and data management rely on predominantly manual practices. In particular, the mediation between different interfaces, data exchange protocols, and differences in meaning in the data made available by systems rests largely on mediators, wrappers, and transformation *pipelines* created manually.

Data sources may exhibit different data quality attributes. Even if the data is structurally and semantically identical, differences in correctness, timelines, and completeness may affect its utility. Moreover, the actual or perceived reliability, quality, or trust may affect the actual or perceived utility.

### 4.2.4 Quality Variety

Data sets may exhibit degrees of data quality, exhibited by various levels of correctness, completeness, and timeliness. Moreover, the actual quality of the data set may differ from the perceived quality if subjective attributes, such as reputation and trust in the sources are taken into consideration. Since the notion of quality is strongly dependent on the application and intended use of the data in a given context, there exists no generally accepted method for quantifying the quality of raw and derived information. Where the quality of data can somehow be assessed directly, for example, by human inspection, comparison to gold standards, or known properties of the sources such as sensors, the data sets can be annotated with quality indicators. Otherwise, metadata describing the lineage of a data set may be used to inform application specific quality assessment methods.

### 4.2.5 Processing Variety

Data processing techniques range from batch- to real-time processing. Batch data processing is an efficient way to process large volumes of data where a series of transactions is accumulated over a period of time. Generally, batch processing uses separate programs for data collection followed by data entry operation and finally the data processing which produce results in batches. Examples of batch processing systems are payroll and billing systems. In contrast, real-time data processing deals with data in (almost) real-time and is suitable for applications where response times should be lower, for example bank automatic teller machines and Point of Sale systems. Sophisticated data processing architectures, such as the Lambda architecture [4], have been developed to address the needs of both batch and real-time processing of large volumes of streamed data.

Differences in processing may further induce varying data quality. The use of multiple software implementations and algorithms for the preprocessing of data, data enrichment, and the extraction of derived data can lead to variations in data quality, which may impede further analysis. It can be difficult to combine and use data sets that suffer from various (systematic) data quality problems arising from, for example, algorithm biases. Comprehensive metadata and provenance information may help to identify any potential issues and point out limitations of the resulting data set.

## 4.3    Variety Management in Data Lakes

The variety of data poses organization challenges that require careful consideration when creating a data lake [5]. If data sets are added in ad-hoc procedures and arbitrary formats, it can be difficult to identify useful data sets, interpret their content, and make effective use of conclusions drawn from joint analysis of multiple data sets. Whether large or small volume, long term maintenance and evolution of the data lake requires thoughtful design and governance to avoid the data lake degenerating into a "data swamp" whose utility declines as more and more data sets are added. Although data lakes are sometimes viewed as *schema-less* data repositories, where any data can easily be added, a minimum of structural and semantic information is required to effectively process and analyze data later – in particular if the users conducting the analysis are different from those having contributed the data sets. The design of data lakes that hold variety-rich data makes use of multi-faceted metadata to facilitate discovery and organization of data sets, federation and partitioning to account for different sources and characteristics of data, and automated enrichment and integration processes that prepare raw data for further analysis while preserving lineage.

### 4.3.1    Metadata Repositories

Metadata about data sets are critical for navigation and discovery of data sets, interpretation of the content of each data set, integration of data, and governance of data lakes. Metadata repositories capture information about entire data sets in addition to more detailed information pertaining to structure and interpretation of the content of each data set.

At the highest level, metadata repositories facilitate navigation and exploration of the contents of a data lake by maintaining an *Information Catalog* based on a domain-specific ontology. Interested readers may refer to Chap. 2 for pragmatic guidance on how to create such an Ontology. The Information Catalog categorizes data sets based on semantic concepts that convey meaning to business users and aid them in exploring and discovering relevant data sets in the data lake. Metadata about the source and timeliness of data can help to assess the quality of data with respect to an intended use.

### 4.3.2    Types of Metadata

Metadata are not limited to categorization of data sets. In general, metadata relevant to the business domain, the technical domain, and the operational domain must be maintained.

- *Business metadata* describes the data in terms that non-technical users can understand. This is particularly important where data lakes are driven by and support non-technical users. Therefore, this vocabulary and classification mechanism should be engineered in close collaboration with business users. The broad categorization of data sets described in the Information Catalog is an example of business metadata.

- *Technical metadata* describes the data type, structure and format of the data. Information about software systems and interfaces required to access and manipulate data are also relevant. Mappings between the business vocabulary and the technical encoding in the data should be maintained to support processing and interpretation of the data. Where possible, a uniform representation should be adopted to simplify later integration of data sets.
- *Operational metadata* captures information about data lineage, temporality, statistical characteristics such as volume, and events related to ingestion, processing, modification, and storage of data. In particular, lineage and temporal information is important for data integration purposes and for assessment of data quality, while information about storage and replication of data supports efficient retrieval. Moreover, restrictions on access and use of data can be captured as metadata. (Chapter 3 outlines an approach where policies are attached to linked data.)

Akin to data schemas describing the information held in a database, a metadata schema should be devised that defines the meaning and representation of metadata. Although the data in a data lake can vary considerably, the metadata representation is typically more uniform and stable.

The quality (especially correctness and completeness) of metadata is critical for operating a data lake. Therefore, tools should be provided to control how this data is generated, where possible to achieve consistency. Technical and operational metadata can often be acquired automatically. For example, lineage and temporality of data as well as data format, time of ingestion, and any processing that may have altered the data can be captured automatically as part of the ingestion and processing pipeline. Business metadata and their mappings to technical representations are often more difficult to obtain. Software tools should be provided that support users in supplying metadata when data is ingested into the lake. Moreover, intelligent assistants that can infer metadata annotations and mappings can simplify acquisition and improve quality.

### 4.3.3   Granularity of Metadata

Metadata can be associated with entire data sets or with individual elements in data sets, depending on the granularity at which data is modified. For data lakes supporting applications such as log file analysis, metadata about entire log files may be sufficient, whereas "data vaults" that operate akin to traditional record-based data stores may require record-level metadata.

### 4.3.4   Federation

Data that is available in a data lake may be stored in external systems, such as relational databases maintained within different organizational boundaries. Federated architectures that retrieve the data from the source systems on demand can leverage the existing systems and provide a single access point for all data. *Wrappers* (also called *adapters*) translate data access from the data lake into queries that the external systems can execute and transform the resulting data into the format used within the data lake. Lineage information and other metadata is also added by the wrapper. Mapping and query transformation technologies exist that can be configured based on the source and target data schemas. Linked data standards, such as RDF, can be used within the data lake to take advantage of linking and graph-based representations, even if the data is stored in a different format. The application architecture described in Sect. 4.4 of this chapter and the mediator-based architectures in Chaps. 3 and 13 are examples of such federated data lake architectures.

### 4.3.5 Partitioning

Data lakes can be partitioned into separate zones based on the lifecycle of the data to accommodate varying characteristics of the data sets. If raw data is kept in a data lake, best practices suggest that this data be kept in a *landing zone* that is separate from the processed data in the lake. Data lakes that are designed for specific analytical purposes are often supported by ingestion pipelines where raw data is processed, cleaned/standardized, enriched, and linked or integrated with other data sets. This process may be separated in different stages where the intermediate results are kept in different zones, and metadata link the related data across zone boundaries. This architecture clearly separates the heterogeneous raw data from structured derived data and enables data lakes to use different technologies and policies to manage the data in each zone. This is beneficial as raw data and processed data typically differ in volume, data type, and access patterns. The links between data in different zones support resolution of problems that can arise in the ingestion pipeline and support additional analysis that may not have been foreseen in the original design of the data lake. Best practices for designing technical architectures for big data processing systems based on heterogeneous technology stacks (so-called "polyglot architectures") can be found in the Big Data Reference Architecture developed by the Data to Decisions Cooperative Research Centre [6].

### 4.3.6 Data Integration and Data Enrichment

Analyzing raw data is often challenging due to differences in representation, quality, and completeness. Moreover, raw data may need to be processed to expose the information required for further analysis, and data sets may need to be integrated with other information in the data lake.

Common operations preceding analysis include discarding invalid or incomplete data, changing the representation (encoding and/or structure, standardization) of data, extracting key information, and creating derived data from a data set. Moreover, algorithms can be applied to enrich data sets by deriving new information by extracting key information from data and linking data to other data sets in the data lake. For example, consider the analysis of unstructured text documents where entities of interest are to be identified in the text and integrated with other information in the data lake. The text documents can initially be stored in a Hadoop cluster, and text can subsequently be extracted and indexed in a free-text search engine such as Elasticsearch[1]. Named entity extraction algorithms can be applied to identify mentions of interesting entities, for example, persons or locations, and a summary of entities can be stored in a structured data store. This data set can form the basis for further analysis, for example entity linking, clustering, and network analysis.

Automated orchestration of pre-processing and enrichment processes has several benefits, including consistency of data processing, automatic maintenance of lineage metadata, and enforcement of data standards within the lake. Moreover, having pre-processed data available in a data lake may enable users with relatively low technology literacy to access the data.

Moreover, defined data standards facilitate the integration of multiple data sets, where each data source is integrated via the common standards rather than each data set being integrated with a number of other data sets via point-to-point integration solutions. Several mediator-based architectures have adopted this approach where (partial) standardization occurs at the mediator. For example, the architecture in Chap. 3 applies semantic lifting at the mediators.

### 4.3.7  Access Control

Access control mechanisms for heterogeneous data lakes can be difficult to implement, as popular big data platforms emphasize scalability over security. Moreover, traditional user- and role-based access control mechanisms may be inadequate for applications where sensitive information is present in the data or in its structure. For example, the access privileges required to see a link between entities in a graph representing relationships between persons may depend not only on the specific link type but also on properties of the entities at the endpoints of the link. Therefore, access policies and access control mechanisms may need to be described at the individual fact level rather than at the data set or even data source level. Although metadata annotations can be furnished to represent the required privileges at appropriate granularity, the tool support for enforcing such fine-grained access models is just emerging. Moreover, enrichment processes and data lineage mechanisms must be carefully designed with security in mind to ensure that appropriate access metadata is associated with derived information. Mediator-based federated architectures face the additional challenge that data sources may be unable to enforce fine-grained access restrictions. Enforcing access control across a heterogeneous software ecosystem is subject to active research.

---

[1] https://www.elastic.co/.

## 4.4    Application in Law Enforcement

Agencies increasingly rely on information that is generated by other agencies or their partners. For example, the D2D CRC works with intelligence agencies and police forces to build an Integrated Law Enforcement (ILE)[2] platform. Government information sources range from criminal histories, immigration records, fingerprint data, gun or vehicle registrations, and video surveillance camera feeds, to house ownership and tax information. However, obtaining information in a timely manner and in a form that is suitable for analysis remains challenging, as analysts may not be aware that data pertaining to their interests is available elsewhere and data is often difficult to obtain across agency and system boundaries. Moreover, access to data must be managed carefully since sensitive content may be subject to legislative constraints and warrants, and replication or import into analytic tools may not be possible.

The goal is to develop, integrate and evaluate technology that will provide police forces and analysts with uniform access to integrated information derived from diverse data sources held by different agencies to aggregate data, resolve and link entities and identities, identify and react to unusual patterns, and build and maintain threat models related to events and entities.

### 4.4.1    Overview of an Integrated Law Enforcement (ILE) Architecture

The overall architecture of the ILE platform is shown in Fig. 4.1. It is based on a federated architecture model where one or more instances of the ILE platform can be deployed and access multiple external data sources. Each instance can provide individual query and analytic services and can obtain data from other instances and external sources on demand. The architecture consists of five layers which are described in more detail below:

**Data Sources:** law enforcement agencies heavily rely on external sources which need to be integrated with internal information systems. External sources are usually controlled by external organizations and may change over time, where change is related to the content, structure and accessibility. Organizational policies in this context rarely support traditional Extract-Transform-Load (ETL) ingestion processes across organizational boundaries, which can be a challenge. External sources can be social media channels like Twitter, Facebook posts or any documents that may be collected during an investigation and are often semi- or unstructured. External services may need to be accessed as well, for example, phone records from telecommunication companies. Internal information systems are systems that help to manage investigations. Depending on the structure of an agency, federal and state-level offices may use different systems or different versions of the same system which adds to the complexity of the integration.

---

[2] http://www.d2dcrc.com.au/rd-programs/integrated-law-enforcement/.

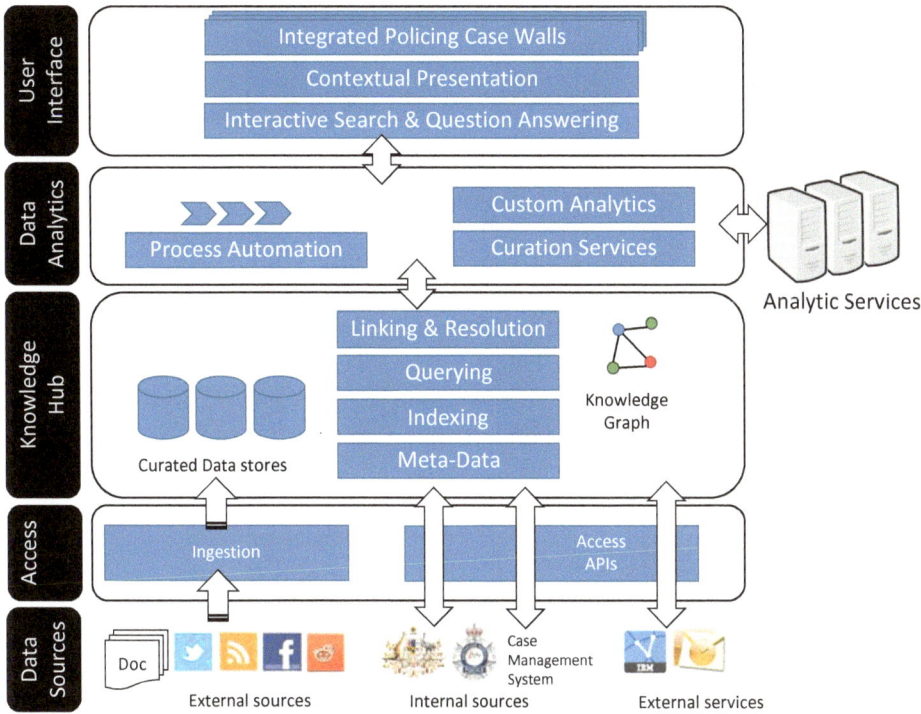

**Fig. 4.1** High-level architecture of Integrated Law Enforcement developed by the D2D CRC

**Access Layer:** the ILE platform provides and uses application programming interfaces (APIs) to front-end applications to access data and invoke analytic services. This two way communication is used where possible, for example, where external services such as phone record querying systems provide an API and internal systems such as an investigation management system calls the ILE API for performing federated queries. The APIs exposed by the platform are using a uniform data format and communication protocol. Mobile applications for investigators may be developed in future versions of the platform that make use of those APIs.

The ingestion subsystem provides access to external sources which do not provide an API, such as documents and some social media channels. For those data sources, the ILE platform provides wrappers. A difference to traditional wrappers and ETL processes is the support for linked-data and support for ingesting metadata instead of loading the complete content of a data source into the platform as done with traditional data warehouse approaches.

**Knowledge Hub:** this layer represents the core of the ILE platform. Data is stored in *Curated Linked Data Stores*, that is, a set of databases that collectively implement a knowledge-graph like structure comprising entities and their links and metadata [7, 8]. This curated data store holds facts and metadata about entities and their links whose veracity has been confirmed. It is used to infer the results for queries and to synthesize requests to external sources and other instances if further information is required. As such, the linked data store implements a directory of entities and links enriched with appropriate metadata including links to an ontology and source information such that detailed information can be obtained from authoritative sources that may be external to the system. This approach is needed as data in the law enforcement domain is dispersed among a number of systems owned and operated by different agencies. As such, no centrally controlled database can feasibly be put in place in the foreseeable future.

The information contained in the linked data store is governed by an ontology that defines the entity types, link types, and associated metadata that is available among the collective platform. The ontology acts as a reference for knowledge management/organization and aids in the integration of information stemming from external sources, where it acts as a reference for linking and translating information into a form suitable for the knowledge hub. The ontology has been designed specifically for the law enforcement domain and includes detailed provenance information and metadata related to information access restrictions. It is explicitly represented and can be queried. All information within the ILE platform is represented in the ontology in order to facilitate entity linking and analysis.

The ontology conceptualizes the domain on three levels: (1) *meta-level* where concept types are captured, (2) the *type level*, where domain concepts are represented in terms of types, and (3) the *instance level*, where instance-level data is represented and linked. For example, the meta-level defines `EntityType`, `RelationshipType`, and `MetaAttributeType`. The types on the level below represent a hierarchy of object types including persons, organizations, vehicles, and other object types, concrete domain relationships that may be established between object types (for example that a Person works for an Organization), and metadata attribute types related to access control, provenance, and temporal validity. Instances at the lowest level represent individual objects and relationships, for example a specific vehicle being registered to an individual person. These domain concepts are closely aligned with the NIEM standard[3] and concepts related to case management. The provenance model is an extension of PROV-O [9]. The instances of the domain concepts form the objects comprising the *Knowledge Graph* on the lowest layer in the ontology. The aforementioned concepts are complemented with classes and objects representing data sources linked to the domain information stored therein as well as schema mapping information required to translate between the external source and the ontology model adopted within the federated architecture. This multi-level modelling method has been adopted to provide a modular and extensible knowledge representation architecture.

Information from external sources is sought based on a catalog of data sources that are available to the system, each with a corresponding adapter that communicates with the external systems and rewrites the information and metadata into the ontology used within

---

[3] https://www.niem.gov/.

the ILE platform [10]. Our platform spans several sources, including an entity database (Person, Objects, Location, Event, and Relations), a case management system, and a repository of unstructured documents.

Information received from external systems is passed through an ingestion and enrichment pipeline where entities are extracted [11], enriched with metadata (provenance and access restrictions) and linked to the knowledge graph in the linked data store. The quality of data is monitored and assured within those pipelines.

**Data Analytics:** analytic services include entity extraction from unstructured text [11], entity linking, similarity calculation and ranking [12]. Services provided by commercial tools, such as network analysis and entity liking/resolution solutions, can be integrated in the modular architecture.

Process Automation services provide workflow orchestration and alert notices if new information relevant to a case becomes available. Workflow services facilitate the enactment of work processes such as acquiring authorization and warrants.

**User Interface:** the ILE platform provides access to data and analytics services through an API (note that APIs are only explicitly shown in the Access Layer in Fig. 4.1) which allows various user interfaces to connect to the platform. Three independent user interfaces are being implemented: (1) Case Walls [13] are a Facebook-like user interface that allows end users to manage investigations in a simple intuitive way that does not require much training, (2) simple federated query forms based on React[4] which can be integrated easily into websites, and (3) a new state-of-the-art virtual environment that allows end users to interact with and explore data of investigations and court cases in a state-of-the-art virtual environment tailored specifically to handling the queries and responses made possible by this system, allowing end users to interact with and explore data of investigations and court cases in novel and innovative ways [14, 15].

Cross-cutting technical concerns, including access control and user management, logging, monitoring and other deployment facilities, have been omitted in this architecture view. Our implementation builds on open source big data technologies, including Hadoop, Spark, PostgreSQL, RabbitMQ, and RESTful interfaces.

## 4.4.2 Addressing Variety Management

The ILE architecture described previously is aimed at addressing and managing the different varieties in the context of big data.

**Structural Variety and Source Variety:** Access to varying data types and data formats is provided through wrappers that are either part of the ingestion in the access layer or translate the payload of API calls. The development of wrappers can be facilitated through model-driven techniques such as model transformation languages, e.g., ATL [16] or ETL

---

[4] https://reactjs.org/.

[17]. These languages allow the lifting of data specifications to a model level, making them easier to understand and manage, and then performs transformations on them to overcome differences in the specifications. The development of transformation rules can be supported by semantic matching technologies such as ontology matching [18] and matching tools such as Karma [19] to overcome semantic variety.

**Granularity Variety:** Managing data on multiple levels of aggregation can be addressed on the access layer and on the data analytics level. On the access layer, wrappers can take care of aggregating data through transformations if the data provided by external sources is too detailed for the ingestion into the curated linked data store. On the data analytics layer, various services can aggregate data from the knowledge hub through operations similar to a data warehouse or OLAP operation if required by the analysis.

**Processing Variety:** Various data processing techniques can be executed within the ILE platform through data analytics services which can access the knowledge hub through an API. The API provides access to the actual data but also allows the querying of metadata and provenance information that can help to identify limitations of resulting data sets.

**Quality Variety:** Data quality can be addressed on the access layer and on the data analytics layer. During the ingestion process or when querying APIs of external services and data sources, data quality can be monitored based on rules executed within wrappers. On the data analytics layer, special data quality services may be applied that analyse the outcomes of analytics services in combination with metadata and provenance data to measure the quality and provide feedback on the result.

## 4.5    Recommendations

Based on our experiences dealing with data lakes and automated data processing systems, the following recommendations are considered critical factors for the long term success of data lakes:

Determine the stakeholders and their intended use cases for the data lake. Their requirements inform the development of the overall architecture and ontologies for data and metadata.

Associate comprehensive metadata with each data set (if possible also with elements in data sets). Automate as much of the metadata acquisition as possible.

Decide on the level of standardization within the lake and implement appropriate ingestion and access channels. Federated access and virtual schemas can provide the illusion of a uniform data lake on top of heterogeneous systems.

Anticipate that the ontologies and user requirements will evolve. Avoid fixing key assumptions about ontologies, data sets, and sources in the implementation.

Provide tools for end users to populate, navigate, and explore the data lake.

Further recommendations and checklists are provided in [5, 20].

## 4.6   Summary

Here, we summarized characteristics of heterogeneous data lakes and investigated the forms of variety that may need to be addressed when collecting a vast variety of data in a data lake. Data lakes typically collate vast amounts of heterogeneous data held in many systems and contributed by a diverse user group. In this context, comprehensive metadata management is essential to long-term maintenance of data lakes and avoiding deterioration of the lake over time.

We summarized types of data heterogeneity and discussed the importance of metadata and ontologies for organizing data lakes. Here, ontologies can aid discovery, navigation, exploration, and interpretation of heterogeneous data lakes. In addition to providing a domain-specific vocabulary for annotation, retrieval, and exploration of data sets by domain experts and data analysts, ontologies can help add a lightweight *schema* on top of the data lake. Virtualization techniques, such as wrappers and mediator architectures, can help overcome structural variety in data and associate data with critical metadata through automated ingestion and data access processes. Semantic metadata is a cornerstone for such architectures, as it describes structure, provenance, visibility (access control) and (permitted) use. As such, ontologies and comprehensive metadata catalogs can simplify interpretation, lift data quality, and simplify integration of multiple data sets. However, appropriate governance mechanisms for ontology evolution are required to sustain data quality, in particular if the ontology or ontologies supporting the data lake are not fixed but evolve over time. Therefore, implementing a successful data lake is not simply a technological challenge; related processes, intended use cases, and social factors must be considered carefully.

We presented an application in the law enforcement domain, where diverse data from many sources are to be collected and maintained in stringent evidence gathering processes by police investigators. This application is characterized by data-driven processes, where the future course of action depends to a large degree on the effective exploration and linking of partial information collected in the course of an investigation. The underlying data lake relies heavily on ontologies for mediating between the different information representations in the many information sources and for efficient exploration of the collated information. Metadata captures lineage of data, supports entity linking algorithms, and governs information visibility and access control.

## References

1. Laney D (2001) 3D data management: controlling data volume, velocity and variety. META Group Inc, Stamford, Connecticut
2. NewVantage Partners LLC (2016) Big Data executive survey 2016. NewVantage Partners, Boston, MA

3. Dayley A, Logan D (2015) Organizations will need to tackle three challenges to curb unstructured data glut and neglect. Gartner report G00275931. Updated Jan 2017
4. Marz N, Warren J (2013) Big Data: principles and best practices of scalable realtime data systems. Manning Publications, Manning, New York
5. Russom P (2017) Data lakes: purposes, practices, patterns, and platforms. Technical report, TDWI
6. D2D CRC (2016) Big Data reference architecture, vol 1–4. Data to Decisions Cooperative Research Centre, Adelaide
7. Stumptner M, Mayer W, Grossmann G, Liu J, Li W, Casanovas P, De Koker L, Mendelson D, Watts D, Bainbridge B (2016) An architecture for establishing legal semantic workflows in the context of Integrated Law Enforcement. In: Proceedings of the third workshop on legal knowledge and the semantic web (LK&SW-2016). Co-located with EKAW-2016, ArXiv
8. Mayer W, Stumptner M, Casanovas P, de Koker L (2017) Towards a linked information architecture for integrated law enforcement. In: Proceedings of the workshop on linked democracy: artificial intelligence for democratic innovation (LINKDEM 2017), vol 1897. Co-located with the 26th international joint conference on artificial intelligence (IJCAI 2017), CEUR
9. Lebo T, Sahoo S, McGuinness D, Belhajjame K, Cheney J, Corsar D, Garijo D, Soiland-Reyes S, Zednik S, Zhao J (2013) PROV-O: the PROV ontology. W3C on-line, https://www.w3.org/TR/prov-o/. Last accessed 15 Mar 2018
10. Bellahsene Z, Bonifati A, Rahm E (2011) Schema matching and mapping. Springer, Berlin, Heidelberg
11. Del Corro L, Gemulla R (2013) ClausIE: clause-based open information extraction. In: Proceedings of WWW. ACM New York, NY, USA
12. Beheshti S-M-R, Tabebordbar A, Benatallah B, Nouri R (2017) On automating basic data curation tasks. In: Proceedings of WWW. ACM, Geneva, Switzerland. pp 165–169
13. Sun Y-JJ, Barukh MC, Benatallah B, Beheshti S-M-R (2015) Scalable SaaS-based process customization with CaseWalls. In: Proceedings of ICSOC. LNCS, vol 9435. Springer, Berlin, Heidelberg. pp 218–233
14. Drogemuller A, Cunningham A, Walsh J, Ross W, Thomas B (2017) VRige: exploring social network interactions in immersive virtual environments. In: Proceedings of the international symposium on big data visual analytics (BDVA). IEEE NJ, USA
15. Bastiras J, Thomas BH, Walsh JA, Baumeister J (2017) Combining virtual reality and narrative visualisation to persuade. In: Proceedings of the international symposium on big data visual analytics (BDVA). IEEE NJ, USA
16. Kurtev I, Jouault F, Allilaire F, Bezivin J (2008) ATL: a model transformation tool. Sci Comput Program 72(1):31–39
17. Polack F, Kolovos DS, Paige RF (2008) The Epsilon transformation language. In: Proceedings of ICMT. LNCS, vol 5063. Springer, Berlin, Heidelberg
18. Shvaiko P, Euzenat J (2013) Ontology matching. Springer, Berlin, Heidelberg
19. Szekely P, Knoblock CA, Yang F, Zhu X, Fink EE, Allen R, Goodlander G (2013) Connecting the Smithsonian American Art Museum to the linked data cloud. In: Proceedings of ESWC
20. Russom P (2016) Best practices for data lake management. Technical report, TDWI

# Text Mining in Economics

Melanie Siegel

**Key Statements**
1. Natural Language Processing methods can be used to extract information from economic forecasts and transform it into a structured form for further analysis.
2. An ontology of "Corporate Social Responsibility" information supports automatic indexing and information extraction from management reports.

## 5.1    Introduction

For medium-sized and large corporations located in Germany, the preparation and publication of management reports in addition to annual and consolidated financial statements is mandatory. These documents are available in printed form, PDFs, but also as HTML documents (https://www.bundesanzeiger.de/). They are an important source of information for financial analysts and economists to answer questions such as:

- How has the economic situation developed in the past financial year?
- What kind of developments can be expected in the future?
- Which are the effects of the corporate actions on the environment and society?
- Which measures does a company take with respect to the environment and society?

M. Siegel (✉)
Hochschule Darmstadt, Darmstadt, Germany
e-mail: melanie.siegel@h-da.de

© Springer-Verlag GmbH Germany, part of Springer Nature 2018
T. Hoppe et al. (eds.), *Semantic Applications*,
https://doi.org/10.1007/978-3-662-55433-3_5

However, status reports are not standardized and unstructured.[1] Most of the information is text. The wording is therefore highly variable: the same meaning can be expressed in different sentences. If tables are included, their columns and rows may be named differently in each annual report. The chapter headings may show significant differences between different companies. Even the terminology may be inconsistent. In order to be able to evaluate the large quantities of texts, automatic methods are being sought to support economic science. The automatic methods are intended to support the intellectual analysis with statistical data obtained from the texts. Since the information is text, database technology cannot access the content. A technology is needed that can refer to the semantics of the texts.

In a discussion between an economist (C. Almeling) and a speech technologist (M. Siegel), the idea emerged to investigate which semantic technologies could be applied to business and management reports to support their economic analysis.

Two projects have been carried out to identify Natural language Processing (NLP) techniques that effectively support economic analysts: the analysis of economic forecasts in business and management reports, and the analysis of information on the impact of corporate actions on the environment and society ("Corporate Social Responsibility" CSR) in these reports. Prototypical implementations were done in the Python programming language, using the Natural Language Toolkit (NLTK) and TextBlob packages. The analysis of economic forecasts uses named-entity recognition and pattern-based information extraction. For the analysis of CSR information, NLP techniques were used to build an ontology. On the basis of this ontology, an automatic indexing was implemented. Firstly, however, the text had to be cleaned from markup and enriched with linguistic information.

## 5.2    Preparation of Texts and Analyses Using NLP Methods

The analyses are based on a corpus of management reports of nine companies of different size (Adidas, Allianz, Axel Springer, Daimler, Delticom, Franz Haniel, Hochtief, Merck, United Internet, and Vulcanic Triatherm) from 2012. The status reports of these are published in HTML format. In order to be able to analyse the texts, the HTML markup is cleaned. This was achieved using the Python HTML parser library and the resulting raw text was subsequently cleaned using regular expressions.

The next step is tokenization of the text, i.e. the division of the text into sentences and tokens (words). For this purpose, the NLTK[2]-based libraries TextBlob[2] and TextBlobDE[3] were used.

The morphological analysis of the words to find the basic forms – lemmatisation – is useful to find morphological variants (such as plural and genitive forms) when indexing the text (see also [4]). The simpler process of stemming used by many systems for English is not suitable for the complex German morphological structure, as in stemming endings

---

[1] Language technologists describe texts as unstructured data, as opposed to databases, see [4].

[2] https://textblob.readthedocs.org/en/dev/.

[3] http://textblob-de.readthedocs.org/en/latest/.

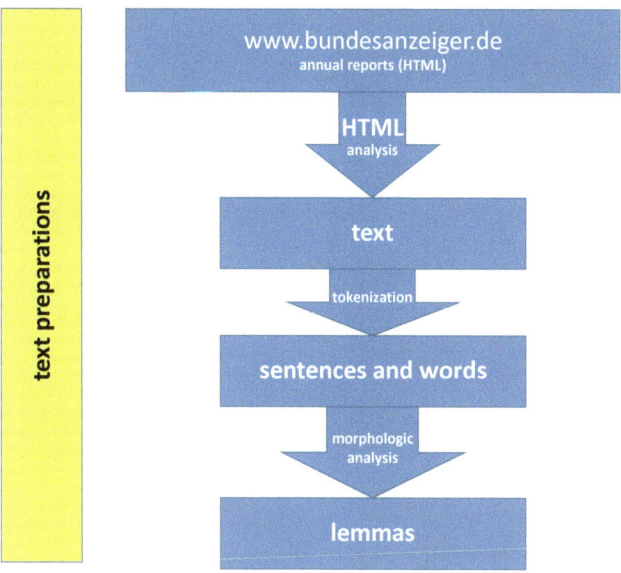

**Fig. 1** Process of text preparations

are just cut off ("beautifully" – "beautiful"). For the domain of annual reports it is necessary to adjust the lemma lexicon for specialized terminology. Therefore, we decided to implement this step ourselves and to use and extend the "German Morphology data based on Morphy" ([5][4]). For the process, see Fig. 1.

## 5.3   Analysis of Economic Forecasts in Annual Reports Using Pattern Rules

The first project was to identify and annotate the texts on economic developments in annual reports in order to gain a structured form of presentation.

Developments and forecasts are expressed in text such as:

- "Infolgedessen rechnen wir mit einer Steigerung des Ergebnisses je Aktie auf einen Wert zwischen 4,25 € und 4,40 €." ("As a result, we expect the earnings per share to increase to between € 4.25 and € 4.40.")
- "Steigende Arbeitslosigkeit, strikte Sparmaßnahmen, geringe Lohnsteigerungen und niedrigere Konsumausgaben werden die Entwicklung des Sporteinzelhandels voraussichtlich negativ beeinflussen." ("Rising unemployment, strict austerity measures, low wage increases, and lower consumption expenditures are expected to have a negative impact on the development of sports retail.")

---

[4] http://www.danielnaber.de/morphologie/.

```
<ANALYSIS>
    <TEXT> sentence with the information</TEXT>
    <FORECAST>
        <ABOUT>type of forecast</ABOUT>
        <ORGANISATION>company or organisation</ORGANIZATION>
        <MARKET>market, e.g. the Asian market</MARKET>
        <DIVISON>department of company or organisation</DIVISON>
        <PRODUCT>product</PRODUCT>
        <VALUE>value</VALUE>
    </ FORECAST>
</ANALYSIS>
```

**Fig. 2** XML format for the structured presentation of forecasts

An XML format (see Fig. 2) is used to represent the information, allowing the storage and retrieval of relevant information in a structured manner.

To fill the ORGANIZATION, DIVISION, PRODUCT, and MARKET fields, a named-entity recognition tool was implemented in Python. The program is based on POS tags[2] that were generated using TextBlob and gazetteer lists of, for example, abbreviations of organizational names (such as "AG", "GmbH") [6]. Pattern rules search for names in the text and provide them for the structured representation.

Expressions for money are also sought with such rules. They work on the basis of numbers and special characters, such as dollar signs.

For the forecast type (in ABOUT), regular expression patterns were used. The following is an example of a pattern that finds an increase announcement:

```
increase_pattern =
re.compile(r'(voraussichtlich|erwartet|rechnen|erwarten|prognostiziere
n|gehen).* .* (Umsatz|Bruttomarge) .* (steigen|ansteigen)')
```

With this information – named entities, money expressions and patterns – the sentences in the annual report are analysed and the fields in the XML structure are filled. An example section of the XML output generated by processing a forecast is given in Fig. 3.

This presentation gives the economist quick access to the parts of the report that are concerned with economic forecasts and to the content of these text parts. XSLT[5] sheets can easily be written for a more ergonomic HTML representation.

---

[5] XSLT is a language that is used to transform XML data to other formats, such as HTML. See https://www.w3schools.com/xml/xsl_intro.asp for more information on this.

```
<ANALYSIS>
        <TEXT>Infolgedessen rechnen wir mit einer Steigerung des Ergebnisses
            je Aktie auf einen Wert zwischen 4,25 € und 4,40 €.
        </TEXT>
        <FORECAST>
                <ABOUT>Aktiensteigerung_auf</ABOUT>
                <ORGANISATION></ORGANISATION>
                <MARKET></MARKET>
                <DIVISON></DIVISON>
                <PRODUCT></PRODUCT>
                <VALUE>einen Wert zwischen 4,25 € und 4,40 €</VALUE>
        </FORECAST>
</ANALYSIS>

<ANALYSIS>
        <TEXT>Steigende Arbeitslosigkeit, strikte Sparmaßnahmen, geringe Lohnsteigerungen
            und niedrigere Konsumausgaben werden die Entwicklung des Sporteinzelhandels
                voraussichtlich negativ beeinflussen.
        </TEXT>
        <FORECAST>
                <ABOUT>Umsatzbeeinträchtigung</ABOUT>
                <ORGANISATION></ORGANISATION>
                <MARKET></MARKET>
                <DIVISON>Sporteinzelhandels</DIVISON>
                <PRODUCT></PRODUCT>
                <VALUE></VALUE>
        </FORECAST>
</ANALYSIS>

<ANALYSIS>
        <TEXT>Für XYZ erwarten wir einen Anstieg des währungsbereinigten Umsatzes
            im mittleren einstelligen Bereich.
        </TEXT>
        <FORECAST>
                <ABOUT>Umsatzsteigerung_um</ABOUT>
                <ORGANISATION>XYZ</ORGANISATION>
                <MARKET></MARKET>
                <DIVISON></DIVISON>
                <PRODUCT></PRODUCT>
                <VALUE>mittleren einstelligen Bereich</VALUE>
        </FORECAST>
</ANALYSIS>
```

**Fig. 3** Example results of forecast analysis (the company name is replaced in the third example, to maintain confidentiality)

## 5.4    Analysis of CSR Information in Business Reports Based on Ontological Information

The second text mining project was to analyse CSR information in business reports.

### 5.4.1    First Data Analysis and Development of the Knowledge Base

The first goal of this part of the project was the search for keywords and discovery of the potential of semantic technologies.

In the first step, automatic terminology extraction was carried out on the texts in the corpus. This extraction is based on components of the software system Acrolinx[6]. Not only terms, but also morphological variants were found, such as "Finanzkennzahl/ Finanz-Kennzahl", "Fälligkeitenstruktur/Fälligkeitsstruktur", and "XETRA-Handel/Xetra®-Handel". A text analysis (Information Retrieval, Information Extraction) should take such variants that occur in texts into account.

Based on this terminology extraction, a list of keywords was intellectually developed by the economics expert and attributed to the indicators of the G4 Guidelines on Sustainability Reporting of the Global Reporting Initiative [3]. On the basis of these keywords, the texts were analysed by language technology tools. In a first step, text mining procedures automatically analysed co-occurrences in order to find additional words that often occur together with the keywords, thereby expanding the list.

An example of co-occurrences for the keyword "Compliance" (with frequency of common occurrence) can be seen in Fig. 4.

Not all co-occurrences identified were good keywords (e.g., "Helmut", "Groß"), but some were worth being considered for the lists.

Another important source of information for expanding the keyword list is an analysis of compound terms. We used linguistic techniques to find compounds with our keywords in the text. We then expanded the word list with these compounds. The example of the keyword "Umwelt" (environment) shows that compounds are found in the texts which an expert for the subject does not necessarily think of (see Fig. 5).

For the construction of the knowledge base, it was necessary to access the lemmatisation of the results, e.g. "Umweltmanagementsysteme" to "Umweltmanagementsystem", but also to find all morphological variants occurring in the texts (e.g. "Umwelt-Managementsystem", "Umweltmanagement-System").

The information obtained by this method was used by the economics expert to draw up a list of the keywords of interest for the analysis. These keywords were organized into clusters that were created after the co-occurrence and compound analyses.

---

[('Risiken', 32), ('Aufsichtsrat', 16), ('Vorstand', 16), ('Groß', 15), ('Konzerns', 14), ('Dr.', 13), ('Allianz', 12), ('Daimler', 12), ('Moderat', 11), ('AG', 11), ('Konzern', 11), ('Mitarbeiter', 10), ('Group', 9), ('Prüfungsausschuss', 8), ('Unternehmens', 8), ('Ausschuss', 8), ('Überwachung', 8), ('Fragen', 7), ('Wirksamkeit', 8), ('Compliance-Risiken', 7), ('Richtlinien', 7), ('Mitglieder', 7), ('Einhaltung', 7), ('Fragen',7), ('Geschäftsentwicklung',7), ('Anteilseignervertreter',7), ('Risikomanagementsystem', 7), ('berichtet', 7), ('Officer', 7), ('Risikomanagement', 7), ('Chief', 7), ('Insurance', 7), ('Aufsichtsrats', 6), ('Kontrollen', 6), ('Rahmen', 6), ('Integrität', 6), ('Perlet', 6), ('Kontrollsystems', 6), ('Risikomanagements', 6), ('Compliance-Organisation', 6), ('Risikomanagementsystems', 6), ('Legal', 6), ('Risk', 6), ('Jahr', 6), ('Helmut', 6), ('Dauer', 5), ('Revisionssystems', 5), ('Entwicklungen', 5)]

**Fig. 4** Co-occurrences of the keyword "Compliance"

---

[6] www.acrolinx.com, [7].

['Auto-Umwelt-Ranking', 'US-Umweltschutzbehörde', 'Umweltangelegenheiten', 'Umweltanstrengungen', 'Umweltaspekte', 'Umweltauswirkungen', 'Umweltbelastung', 'Umweltbereich', 'Umweltbestimmungen', 'Umweltbilanz', 'Umweltdaten', 'Umweltfreundlichkeit', 'Umweltleistung', 'Umweltleistungen', 'Umweltmanagement', 'Umweltmanagementsysteme', 'Umweltnormen', 'Umweltpraktiken', 'Umweltpreis', 'Umweltrichtlinien', 'Umweltrisiken', 'Umweltschonung', 'Umweltschutz', 'Umweltschutzmaßnahmen', 'Umweltschutzrisiken', 'Umweltstandards', 'Umweltstrategie', 'Umweltverantwortung', 'Umweltverfahren', 'Umweltverträglichkeit', 'Umweltwissenschaftler', 'Umweltzeichen', 'Umweltzertifikat', 'Umweltzonen']

**Fig. 5** Compound terms identified for the keyword "Umwelt"

## 5.4.2  Establishing an Ontology

The result of this data analysis was a list of keywords, organized into clusters, and assigned to the indicators of the Global Reporting Initiative (GRI). These indicators are already hierarchically organized in a three-level taxonomy (Fig. 6).

This taxonomy works with multiple inheritances, or rather with multiple relations: the term "Compliance" is in the categories "Produktverantwortung" (product responsibility), "Gesellschaft" (society) and "Ökologie" (ecology). The relation to the categories is not a hyponym but a meronym relation. For this reason, an ontology is needed to organize the knowledge base. Another reason is that a considerable number of multiple relations have been introduced for the further organization of the knowledge base.

The keywords found in the data analysis and organized in clusters were related to the GRI categories using a Python implementation (see Fig. 7).

## 5.4.3  Basic Statistics on CSR Information in Annual Reports

With the resulting knowledge base and the keywords, business reports can now be evaluated regarding various questions. Here, we evaluate the report processing approach in the context of the measurement of sustainability reported in German management reports [1].

### 5.4.3.1  Which Topics Are Covered in the Annual Report? Are the Details Complete?

To answer this question, the text is searched for keywords, sentence by sentence. The keywords are then looked up in the knowledge base to determine whether they belong to one of the subjects to be examined. When multiple keywords occur in a sentence, an attempt is made to associate the set of keywords with a topic area.

The program annotates the sentences with their subject areas and thus gives the economist an indication of where the relevant topics are to be found in the text and whether the required topics are shown in the report (Table 1 shows some examples).

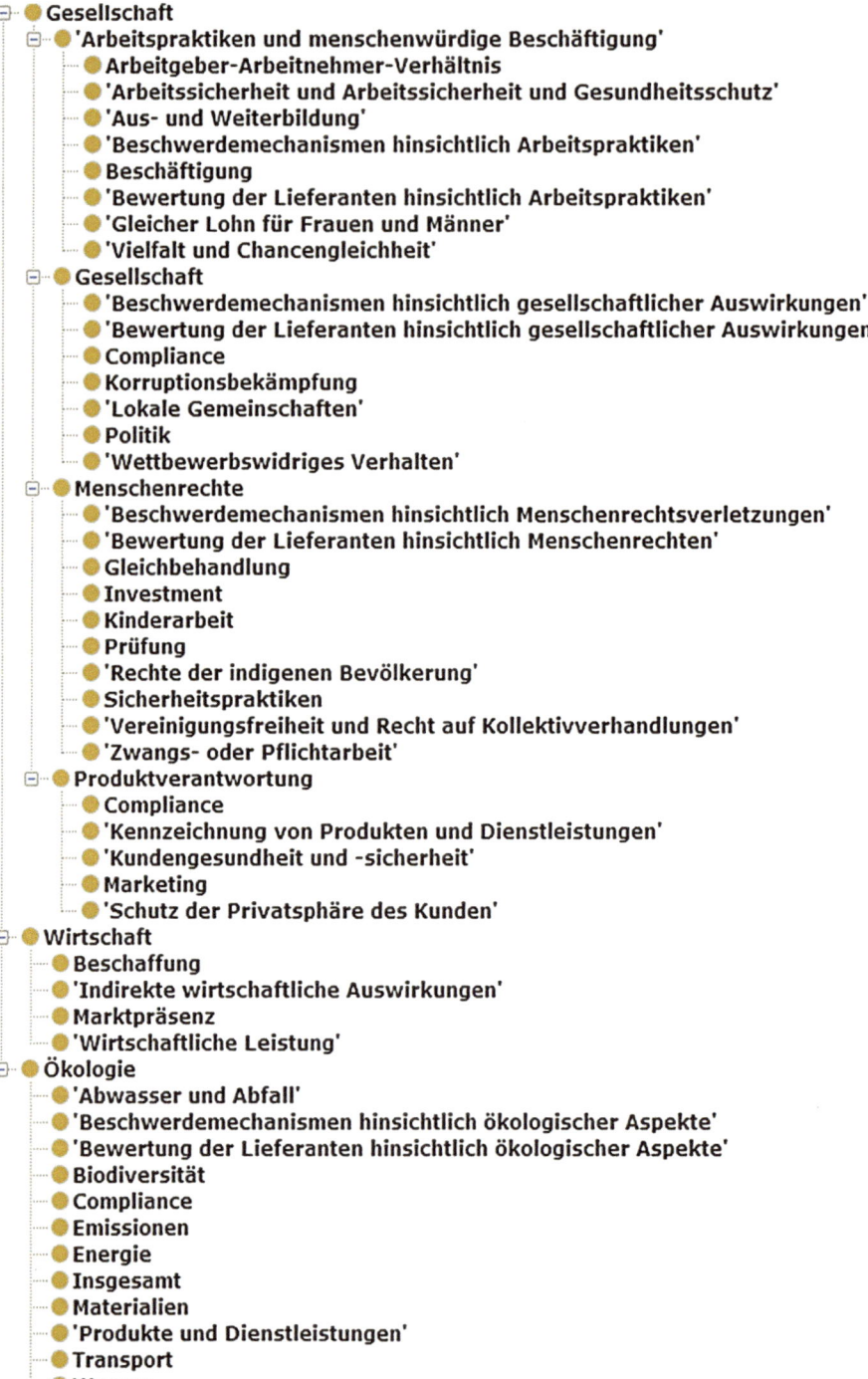

- **Gesellschaft**
  - **'Arbeitspraktiken und menschenwürdige Beschäftigung'**
    - **Arbeitgeber-Arbeitnehmer-Verhältnis**
    - **'Arbeitssicherheit und Arbeitssicherheit und Gesundheitsschutz'**
    - **'Aus- und Weiterbildung'**
    - **'Beschwerdemechanismen hinsichtlich Arbeitspraktiken'**
    - **Beschäftigung**
    - **'Bewertung der Lieferanten hinsichtlich Arbeitspraktiken'**
    - **'Gleicher Lohn für Frauen und Männer'**
    - **'Vielfalt und Chancengleichheit'**
  - **Gesellschaft**
    - **'Beschwerdemechanismen hinsichtlich gesellschaftlicher Auswirkungen'**
    - **'Bewertung der Lieferanten hinsichtlich gesellschaftlicher Auswirkungen'**
    - **Compliance**
    - **Korruptionsbekämpfung**
    - **'Lokale Gemeinschaften'**
    - **Politik**
    - **'Wettbewerbswidriges Verhalten'**
  - **Menschenrechte**
    - **'Beschwerdemechanismen hinsichtlich Menschenrechtsverletzungen'**
    - **'Bewertung der Lieferanten hinsichtlich Menschenrechten'**
    - **Gleichbehandlung**
    - **Investment**
    - **Kinderarbeit**
    - **Prüfung**
    - **'Rechte der indigenen Bevölkerung'**
    - **Sicherheitspraktiken**
    - **'Vereinigungsfreiheit und Recht auf Kollektivverhandlungen'**
    - **'Zwangs- oder Pflichtarbeit'**
  - **Produktverantwortung**
    - **Compliance**
    - **'Kennzeichnung von Produkten und Dienstleistungen'**
    - **'Kundengesundheit und -sicherheit'**
    - **Marketing**
    - **'Schutz der Privatsphäre des Kunden'**
- **Wirtschaft**
  - **Beschaffung**
  - **'Indirekte wirtschaftliche Auswirkungen'**
  - **Marktpräsenz**
  - **'Wirtschaftliche Leistung'**
- **Ökologie**
  - **'Abwasser und Abfall'**
  - **'Beschwerdemechanismen hinsichtlich ökologischer Aspekte'**
  - **'Bewertung der Lieferanten hinsichtlich ökologischer Aspekte'**
  - **Biodiversität**
  - **Compliance**
  - **Emissionen**
  - **Energie**
  - **Insgesamt**
  - **Materialien**
  - **'Produkte und Dienstleistungen'**
  - **Transport**
  - **Wasser**

**Fig. 6** Taxonomy value creation of GRI

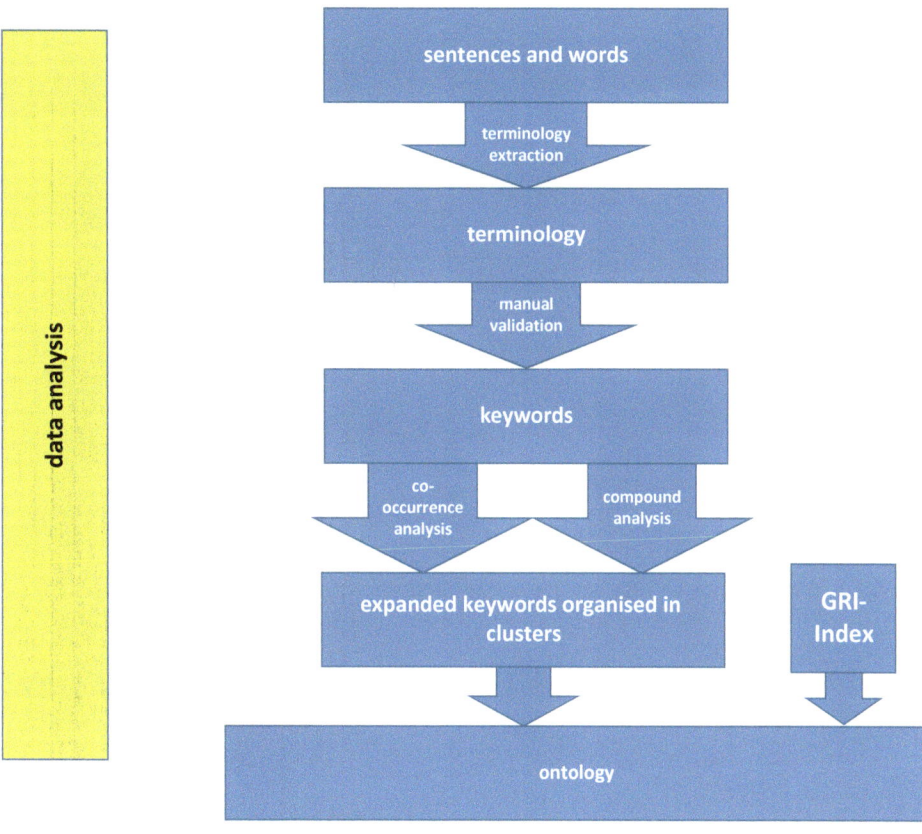

**Fig. 7** Data analysis and establishment of the ontology

**Table 1** Topic annotation example

| Subject area | Text examples |
| --- | --- |
| Ökologie_Compliance *(ecology compliance)* | Bis Ende 2012 konnten zehn Standorte des Konzerns diese Zertifizierung erreichen. *(By the end of 2012, ten of the corporation's sites had achieved this certification.)* |
| Ökologie_Emissionen *(ecology emissions)* | Insgesamt möchten wir bis zum Jahr 2015 die relativen $CO_2$-Emissionen an unseren eigenen Standorten um 30% reduzieren. *(Overall, we want to reduce relative $CO_2$ emissions at our own sites by 30% by 2015.)* |
| Ökologie_Emissionen *(ecology emissions)* | Darüber hinaus wurden zahlreiche andere Maßnahmen umgesetzt, um die $CO_2$-Bilanz der Beschaffungs- und Vertriebskette zu verbessern. *(In addition, numerous other measures have been implemented to improve the carbon footprint of the supply and distribution chain.)* |

| | Absolute | Percentage |
|---|---|---|
| sentences in text | 4086 | |
| sentences with CSR words | 1262 | 30.89 % |
| society sentences | 943 | 23.08 % |
| economy sentences | 363 | 8.89 % |
| ecology sentences | 609 | 14.90 % |
| sentences with other CSR words | 475 | 11.63 % |
| words in text | 83092 | |
| CSR words in text | 1853 | 2.23 % |
| society words | 1304 | 1.57 % |
| economy words | 440 | 0.53 % |
| ecology words | 758 | 0.91 % |
| other CSR words | 570 | 0.69 % |

**Fig. 8** Example for statistical text analysis

### 5.4.3.2 What is the Extent of the Reporting on the Environment, Society and the Economy in Business Reports in Relation to the Overall Text?

To answer this question, the number of sentences and keywords devoted to the relevant topics were examined, and the corresponding percentage of the total text was determined.

Figure 8 shows an example output.

### 5.4.3.3 How Has the Topic Area Developed in Business and Management Reports Over Time? How Does the Subject Treatment Compare in Different Industries?

The statistical text analysis information can be used to compare the amount of text concerning CSR in the annual reports of different companies. Furthermore, annual reports can be analysed over several years to see how the share of the CSR text changes. In this way, it is possible to examine how much value the companies place on the subjects under investigation.

## 5.5 Recommendations

1. Complex information extraction tasks can benefit from a carefully set up domain ontology that combines concepts and lexical items.
2. An analysis of texts in the domain is useful to organize the ontology and find relevant words and concepts. Methods of natural language processing are of extreme value for this purpose.

## 5.6  Summary

Regarding the question "Can NLP methods effectively support economists in their analyses?", we have made prototypical implementations of NLP systems to analyse the status reports of different companies. As is typical for NLP approaches, as the initial step, the texts are cleaned up and tokenized. Then, the tokens (words) need to be enriched with linguistic information.

In a first project, we analysed the economic forecasts in the texts, extracted information from them, and transformed it into a structured form. Economists thus gained quick access to the text parts with the relevant information. Even this information extraction is thus valuable on its own.

A second project focused on CSR information in the texts. In order to effectively support the answer to three economic questions, we first set up an ontology. For the development of this ontology, various NLP techniques were applied, such as terminology extraction, co-occurrence analysis, and compound analysis. Using the ontology, we were able to derive basic descriptive statistics about value-added information in the annual reports that can be used to answer questions relating to specific topics, such as CSR.

The implementations are hitherto prototypical and not yet complete. However, we have already been able to identify text analysis techniques that can effectively support economic analysts. It is now necessary to further develop and adjust these techniques.

Since there is, so far, no data annotated for this task, a systematic evaluation of the results could not be carried out. This will be an important next step. Once annotated data are available, machine learning methods for automatic indexing can also be tested.

Another important next step would be the implementation of a GUI that allows economists to easily access NLP techniques and analyse texts.

## References

1. Almeling C (2016) Messung öffentlicher Wertschöpfung. Folien zum Brown Bag Seminar am 13.01.2016, Hochschule Darmstadt
2. Bird S, Garrette D, Korobov M, Ljunglöf P, Neergaard MM, Nothman J (2012) Natural language toolkit: NLTK, version 2
3. Global Reporting Initiative (2015) G4 Leitlinien zur Nachhaltigkeitsberichterstattung – Berichterstattungsgrundsätze und Standardangaben
4. Heyer G, Quasthoff U, Wittig T (2006) Text Mining: Wissensrohstoff Text–Konzepte. Algorithmen, Ergebnisse. W3L-Verlag, Bochum
5. Lezius W, Rapp R, Wettler M (1998) A morphological analyzer, disambiguator and context-sensitive lemmatizer for German. In: Proceedings of the COLING-ACL 1998, Montreal
6. Morik K, Jung A, Weckwerth J, Rötner S, Hess S, Buschjäger S, Pfahler L (2015) Untersuchungen zur Analyse von deutschsprachigen Textdaten. Technische Universität Dortmund, Tech. Rep. 2, 2015
7. Siegel M, Drewer P (2012) Terminologieextraktion – multilingual, semantisch und mehrfach verwendbar. In: Tagungsband der TEKOM-Frühjahrstagung 2012, Karlsruhe

# Generating Natural Language Texts

**6**

## Hermann Bense, Ulrich Schade, and Michael Dembach

**Key Statements**
1. Natural Language Generation (NLG) allows the generation of natural language texts to present data from a structured form.
2. NLG is a founding technology for a new sector in the publishing industry.
3. Natural language texts generated automatically are of medium quality with respect to cohesion and coherence. Their quality can be improved, however, by applying methods from the cognitive process of language production, e.g. the mechanisms determining sequencing of phrases which exploit information structure in general and the semantic roles of the phrases, in particular.
4. The number of automatically generated news articles will exceed those written by human editors in the near future since NLG allows hyper-personalization, the next key factors for reader contentment.

## 6.1 Introduction

The publishing industry has a rapidly increasing demand for unique and highly up-to-date news articles. There are huge amounts of data continuously being produced in the domains of weather, finance, sports, events, traffic, and products. However, there are not enough

H. Bense (✉)
TextOmatic AG, Dortmund, Germany
e-mail: hermann.bense@textomatic.ag

U. Schade · M. Dembach
Fraunhofer-Institut FKIE, Wachtberg, Germany
e-mail: ulrich.schade@fkie.fraunhofer.de; michael.dembach@fkie.fraunhofer.de

© Springer-Verlag GmbH Germany, part of Springer Nature 2018
T. Hoppe et al. (eds.), *Semantic Applications*,
https://doi.org/10.1007/978-3-662-55433-3_6

75

human editors to write all the stories buried in these data. As a result, the automation of text writing is required. In Bense and Schade [2], we presented a Natural Language Generation (NLG) approach which is used for automatically generating texts. In these texts, data that is available in structured form such as tables and charts are expressed. Typical examples are reports about facts from the mentioned domains, such as weather reports.

Currently, texts generated automatically are correct, but of medium quality and sometimes monotonous. In order to improve the quality, it is necessary to recognise how semantics in general and information structure in particular are implemented in good texts. In order to illustrate this we need to compare the NLG approach to the cognitive process of language production. Therefore, Sect. 6.3 sketches the cognitive process. In Sect. 6.4, we will discuss what kinds of texts can be successfully generated automatically. Some technical aspects, especially those by which background knowledge can be exploited for the generation, will be presented in Sect. 6.5. In Sect. 6.6, we discuss methods we are currently developing to further enhance the quality of generated texts in terms of cohesion and coherence. These methods exploit insight from the cognitive process of language production as well as the linguistic theory of "topological fields". We summarise our results in Sect. 6.7 and give an outlook to hyper-personalization of news, the next trend in NLG.

## 6.2    The Cognitive Process of Language Production

In 1989, Prof. Dr. Willem J.M. Levelt, founding director of the Max Planck Institute for Psycholinguistics in Nijmegen, published "Speaking: From Intention to Articulation" [11]. This influential monograph merged the knowledge from multiple insights on the cognitive process of language production into one consistent model. Levelt's model is based on models by Karl Bühler [4], Victoria Fromkin [6], Merrill Garrett [7], and J. Kathrin Bock [3], and includes insights on monitoring and error repair elaborated by Levelt himself [10]. It incorporates important advancements about the structure of the mental lexicon [9] and the process of grammatical encoding [8] by Gerard Kempen, as well as equally important advancements about the process of phonological encoding by Gary S. Dell [5]. To this day, Levelt's model provides the base for research on language processing. Levelt, together with his co-workers (among them Antje S. Meyer, Levelt's successor as director at the MPI in Nijmegen, Ardi Roelofs, and Herbert Schriefers) contributed to that research by examining the subprocess of lexical access, see for example Schriefers, Meyer and Levelt [15] and Levelt, Roelofs and Meyer [12].

To compare NLG approaches to the cognitive process of language production, it is of specific importance to take a look at Levelt's breakdown of language production into subprocesses, a classification that is still widely accepted in the field. Levelt distinguishes preverbal conceptualization, divided into macro planning and micro planning, from the linguistic (and thus language dependent) processes of formulation, divided into grammatical encoding and phonological encoding, and articulation, the motoric subprocesses of speaking (and writing). Speaking (and of course also writing) is triggered by an intention. The speaker acts by speaking in order to inform the listener about something, to manipulate the listener to do something, or to convince the listener that he or she will do something. Conceptualization in general, and macro planning in particular, starts with that intention. Considering the intention, the process of macro planning determines the content of the next part of an utterance, i.e., the content of the next sentence. To do so, macro planning exploits different kinds of knowledge at the disposal of the speaker. This knowledge includes encyclopaedic knowledge, e.g., that Robert Lewandowski is a Polish star striker (encyclopaedic knowledge about soccer) and discourse knowledge, e.g., what had already been mentioned, who is the listener, what is the background of the dialog and more. Micro planning takes the determined contents and compresses it into a propositional structure which is coined the "preverbal message" by Levelt. According to Levelt, the preverbal message still is independent from language.

The transformation of the preverbal message into the target language is the task of formulation, the second major subprocess of language production. First, for each concept which is part of the preverbal message, a lexical entry is determined. For example, the concept of a share may trigger lexical entries like "share" or "stock". A competition process then decides whether "share" or "stock" will be used in the resulting expression. The selected entries will be expanded into corresponding phrases, e.g. "the share", in parallel. To achieve this, a procedure inspects the preverbal message in order to determine the specific forms of those phrases. For example, in a noun phrase, a decision has to be made as to whether a determiner is needed and if so, whether the determiner has to be definite or indefinite, whether the noun is singular or plural, and whether additional information has to be incorporated, e.g. in the form of adjectives. In some cases, the noun phrase can even be expressed in the form of a single personal pronoun. Starting with the first concept for which the corresponding phrase is completed, formulation's subprocess of grammatical encoding starts to construct a sentence in which all the phrases are integrated. Of course the concept that represents the action in the message is transformed not into a phrase but into the sentence's verb group. In order to execute the process of grammatical encoding, speakers use all their knowledge about the target language, their vocabulary and their grammatical expertise.

The result of grammatical encoding can be seen as a phrase structure tree, with words as terminals. The representations of these words (lemmata) become the subject of the formulation's second subprocess, which in case of speaking is coined phonological encoding. This process transforms the words into their sequence of phonemes (or letters in the case of writing). Phonological encoding taps into the speaker's knowledge about how to pronounce (or spell) a word. Finally, the articulation process takes over and generates overt speech (or written text).

## 6.3    Automated Text Generation in Use

The main areas for automated text generation are news production in the media industry, product descriptions for online shops, business intelligence reports and unique text production for search engine optimization (SEO). In the field of news production vast amounts of data are available for weather, finance, events, traffic and sports. By combining methods of big data analysis and artificial intelligence, not only pure facts are transferred into readable text, but also correlations are highlighted.

A major example is focus.de, one of the biggest German online news portals. They publish around 30,000 automated weather reports with 3 days forecast for each German city each day. Another example for high-speed and high-volume journalism is handels-blatt.com. Based on the data of the German stock exchange, stock reports are generated for the DAX, MDax, SDax and TecDax indexes every 15 minutes. These reports contain information on share price developments and correlate it to past data such as all time highs/lows, as well as to data of other shares in the same business sector.

An important side effect resulting from publishing such big numbers of highly relevant and up-to-date news is a considerably increased visibility within search engines such as Google, Bing etc. As a consequence, media outlets profit from more page views and revenues from affiliate marketing programs.

From the numbers of published reports it is clear that human editors are not able to write them in the available time. In contrast, automated text generation produces such reports in fractions of a second, and running the text generation tools in cloud based environments adds arbitrary scalability since the majority of the reports can be generated in parallel. Thus, in the foreseeable future, the amount of generated news will exceed that of news written by human authors.

## 6.4    Advanced Methods for Text Generation

In this section we will sketch a semantic approach to augment our generation approach. The base functionality of our tool, Text Composing Language (TCL), used for text generation has already been described in Bense and Schade [2]. In short, TCL is a programming language for the purpose of generating natural language texts. A TCL program is called a template. A template can have output sections and TCL statements in double square brackets. The eval-statement enables calls to other templates as subroutines.

The semantic expansion we want to discuss here aims at adding background knowledge as provided by an ontology. This corresponds to the exploitation of encyclopaedic knowledge by the cognitive "macro planning". The ontological knowledge for TCL is stored in a RDF-triple store[1] which has been implemented in MySQL. The data can be accessed via query interfaces on three different layers of abstraction. The top most layer provides a kind

---

[1] https://en.wikipedia.org/wiki/Triplestore.

of description logic querying. The middle layer, OQL (Ontology Query Language), supports a query interface which is optimized for the RDF-triple store. OQL queries can be directly translated into MySQL-queries. Triples are of the form (s, p, o), where s stands for subject, p for property and o for object. The basic OQL-statements for the retrieval of knowledge are getObjects (s, p) and getSubjects (p, o), e.g. getObjects ('>Pablo_Picasso', '*') would retrieve all data and object properties of the painter Pablo Picasso, and getSubjects ('.PlaceOfBirth', 'Malaga') would return the list of all subjects, who were born in *Malaga*. According to the naming conventions proposed in Bense [1], all identifiers of instances begin with the >-character, those for classes with the ^-character, data properties with a dot and names of objects properties start with <>. TCL supports knowledge base access via the get(s,p,o) function. Depending on which parameters are passed, internally either *getObjects* or *getSubjects* is executed, e.g. getSubjects ('.PlaceOfBirth', 'Malaga') is equivalent to get ('*', '.PlaceOfBirth', 'Malaga'). An example for a small TCL program is:

```
[[ LN = get ('>Pablo_Picasso', '.LastName', '')]]
[[ PoB = get ('>Pablo_Picasso', '.PlaceOfBirth, '')]]
$LN$ was born in $PoB$.
```

This TCL program creates the output: "Picasso was born in Malaga".

The graph in Fig. 6.1 shows an excerpt of the knowledge base about a soccer game. Instances are displayed as rounded rectangles with the IDs of the instances having a dark green background colour [1]. The data properties are shown as pairs of attribute names and their values. The named edges which connect instance nodes with each other represent the object properties (relationship types) between the instances, e.g., <>is_EventAction_of and <>is_MatchPlayerHome_of. The schema behind the example data contains classes for ^Teams ('T_'), ^MatchFacts ('MF_'), ^MatchEvents ('ME_'), ^Player ('P_'), ^Match_PlayerInfo ('MP_P_'), ^Stadium ('STD') and ^City ('CIT'). The match is connected to its teams via (>MF_160465, <>HomeTeam, >T_10) and (>MF_160465, <>AwayTeam, >T_18). All match events are aggregated to the match by the object property <>is_EventAction_of. The inverse of <>is_EventAction_of is <>EventAction. An event action has a player associated with an ^Match_PlayerInfo instance, e.g., by <>MatchPlayerScore in the case the player scoring a goal, or an assisting player is connected to the event using the object property <>AssistPlayer. Each ^Match_PlayerInfo instance is associated with a player via the <>Player relationship type. Finally, each team has a stadium (<>Stadion, inverse object property: <>ist_Stadion_von) and each stadium has an associated city (<>ORT).

The data model behind the application for the generation of premier league soccer match reports is much more complex, but the small excerpt gives a good impression of the complexity it deals with. Accessing the information needed for generating text output for a report can be a cumbersome task even for experienced database programmers. The following explains the implementation of a method that can quickly retrieve information out of these graphs. In principle, the terms sought can be easily derived even by non-programmers, by following the path from one instance in the knowledge graph to the

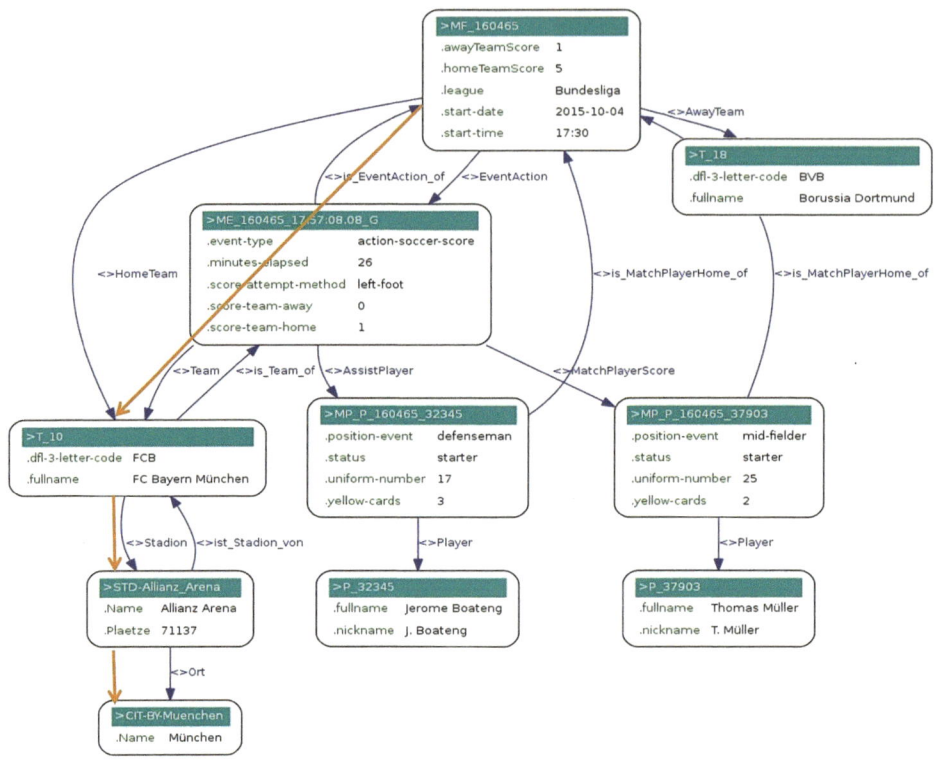

**Fig. 6.1** Ontological knowledge to be exploited for generation of soccer reports

targeted instance, where the needed information is stored. The path (the orange arrows in Fig. 6.1) starting at the instance node of the match **>MF_160465<>Hometeam<>Stadion<>Ort.Name** follows the chain of properties **<>Hometeam<>Stadion<>Ort.Name** to give access to the name of the city where the event takes place. A property chain is the concatenation of an arbitrary number of object property names, which can be optionally followed by one data property name, in this case **.Name**.

In TCL, templates can be evaluated on result sets of OQL queries. The query getObjects ('>MF_160465', '<>HomeTeam') positions the database cursor on the corresponding triple of the knowledge base. In a template, the values of the triple can be referenced by the term **$S$**. Beyond this, the TCL runtime system is able to interpret property chains on the fly. Therefore it is possible to have the following declarations as part of a template header:

```
STRT = $S.start-time$
DTE = $S.start-date;date(m/d/Y)$   /* formatted in English date format
STDN = $S<>HomeTeam<>Stadion.Name$
CTYN = $S<>HomeTeam<>Stadion<>Ort.Name$
```

Then the template *"The game started on $DTE$ at $STRT$ o'clock in the $STDN$ in $CTYN$."* generates for the sample data the output *"The game started on 10/4/2015 at 17:30 o'clock in the Allianz Arena in München."*.

Internally, an automatic query optimization is applied for property chains. The processing of property chains is an iterative process, where initially the subject is retrieved together with its first property. The resulting object becomes the new subject, which is then retrieved in combination with the second property and so forth. Each retrieval is realized by an SQL-SELECT. The length of the property chain determines, how many queries have to be executed. Therefore, starting from the match instance, four queries are needed to retrieve the name of the city where the match takes place. The query optimizer takes the complete property chain and internally generates and executes a nested SQL-Query. Performance benchmarks have shown that when property chains are used, execution time can be significantly reduced.

## 6.5   Increasing Quality by Exploiting Information Structure

In this section, we will discuss how to increase the quality of the generated texts by exploiting the semantic principle of information structure. The selection of another lexical entry for a second denotation of a just mentioned concept (e.g., in order to denote a share, the term "stock" can be used in English; in German "Wertpapier" can substitute "Aktie") increase readability and text quality. In the cognitive process of lexical access, this principle is incorporated naturally as used items are set back in activation and have to recover to show up again. Sometimes, the same holds for grammatical patterns: consecutive SPO sentences feel monotonous. We will discuss an approach to automatically vary sentence patterns below. With this approach, we make available a set of grammatical patterns that can be used to generate the next expression. Having this set available, we can prune it semantically to emulate information structure. In order to clarify what is meant by "information structure" from the perspective of the cognitive process, we will shortly discuss its lexical counterpart. In the Levelt model, the concepts of the preverbal message are annotated according to their "availability" (whether they have already been mentioned before). This might lead to the selection of a different lexical entry as discussed. Alternatively, complex nouns in noun phrases can be reduced to their head ("Papier" instead of "Wertpapier"). Noun phrases even can be reduced to the corresponding personal pronoun, if the respective concept is in "situational focus". For example, "Robert Lewandowski has been put on in minute 62. Robert Lewandowski then scored the goal to 2-1 in minute 65" can and should be substituted by "Robert Lewandowski has been put on in minute 62. He then scored the goal to 2-1 in minute 65" in order to generate a cohesive text. In Bense and Schade [2], we already discussed an algorithm that can handle these kinds of cases. In addition to this, noun phrases that are constituted by a name and that in principle can be reduced to a pronoun, can also be substituted by another noun phrase that expresses encyclopaedic knowledge. Considering again the "Robert Lewandowski" example, the second occurence of his name in the original text could be substituted with "The Polish national player" which adds information and makes the whole expression more coherent [14].

We also developed a program that generates the possible variations for given sentences. In doing so we make use of the fact that, especially in German and English, the word order is determined by certain rules and structures. Phrases have already been introduced. There are a few tests at hand which help clarify whether a certain group of words form a phrase or not. One of these tests, the permutation test, checks if the words in question can be moved only as a whole. In example (2) the word sequence "in the 65th minute" is moved. The result is a correct sentence, so the sequence is a phrase. In (3) only "65th minute" is moved. The result is not grammatically correct as indicated by an *. Thus, "65th minute" is not a phrase on its own.

(1) Lewandowski scored in the 65th minute.
(2) In the 65th minute, Lewandowski scored.
(3) *65th minute Lewandowski scored in the.

The mobile property of phrases is used to determine the variants of a sentence, but in order to do so, another linguistic concept must be taken into consideration.

For practical reasons, the German language is the most important to us and its word order can be described quite conveniently with so called topological fields (a good description can be found in Wöllstein [16]. Similar approaches hold for most other Germanic languages, e.g., Danish, but not for English. The topological field approach separates a sentence into different fields corresponding to certain properties. Three basic types are distinguished using the position of the finite verb as the distinctive characteristic. The types are illustrated by the three sentences in Table 6.1. In V1-Sentences, the finite verb is the first word of the sentence and builds the so called Linke Klammer (left bracket), which – together with an optional Rechte Klammer (right bracket), built by the infinite part of a complex predicate – surrounds the Mittelfeld (middle field; contains all the other parts of the sentence). This type of sentence corresponds mostly with the structure of questions. In the case of V2-Sentences, the finite verb is preceded by exactly one phrase – the verb therefore occupies the second

**Table 6.1** The German sentence types illustrated by examples – the example sentences translate to "Did Lewandowski run 100 Meter?", "Lewandowski ran 100 meter because…" and "… because Lewandowski ran 100 meter", respectively

| Type | Vorfeld (prefield) | Linke Klammer (left bracket) | Mittelfeld (middle field) | Rechte Klammer (right bracket) | Nachfeld (final field) |
|---|---|---|---|---|---|
| V1 | | Ist | Lewandowski 100 Meter | gelaufen? | |
| V2 | Lewandowski | ist | 100 Meter | gelaufen, | weil… |
| VL | | weil | Lewandowski 100 Meter | gelaufen ist. | |

spot – in the Vorfeld (prefield). The rest of the sentence is the same as the V1-Sentence, except for the addition of the Nachfeld (final field), which can be found after the Rechte Klammer (right bracket) and mostly contains subordinate clauses. This type mostly corresponds with declarative sentences. Finally, there is the VL-Sentence (verb-last sentence) and as the verb is in last position, its construction is a bit different. Vor- and Nachfeld are not occupied, and a subjunction fills the Linke Klammer while the whole predicate is in the Rechte Klammer. The Mittelfeld is again filled with the rest of the sentence.

The different properties of the fields are numerous enough to fill several books. Here, two examples shall be sufficient to demonstrate in which way we make use of which principles. We will focus on V2-Sentences because their relative abundance makes them the most important sentences to us. The sentences (4)–(6) all have the same proposition – that yesterday Lewandowski ran 100 meter – but we will not translate all the variations being discussed, because rules of word order can be very language specific, i.e. one word order might be wrong in two languages for two different reasons. Trying to imitate the different word orders of German might therefore lead to false analogies in the reader's perception. A prominent property of the Vorfeld – and an important difference to its English equivalent – is its limitation to only one phrase. The following sentence, which has an additional "gestern" ("yesterday") is incorrect because two phrases occupy the Vorfeld:

(4)  *Lewandowski gestern ist 100 Meter gelaufen.

The properties of the Mittelfeld mostly concern the order of its phrases. The subject – in this example "Lewandowski" – is mostly the first element in the Mittelfeld, if it doesn't occur in the Vorfeld already. Therefore, example (6) is grammatically questionable as indicated by a '?' while (5) is correct.

(5)  Gestern ist Lewandowski 100 Meter gelaufen.
(6)  ?Gestern ist 100 Meter Lewandowski gelaufen.

Interesting to us is the fact that the limitation of the Vorfeld actually concerns the concept of the phrase and not just a few select words. The following sentence is absolutely correct in German:

(7)  Der in Warschau geborene und bei Bayern München unter Vertrag stehende Fußballspieler
     Robert Lewandowski ist gestern nur 100 Meter gelaufen. ("The soccer player Robert
     Lewandowski, who was born in Warsaw and is under contract at Bayern München, has
     run 100 meters yesterday.")

This shows that phrases and topological fields are not just concepts invented by linguists in order to describe certain features of language more accurately, but reflect actual rules, which are acquired in some form and used during speech production. Therefore, we want to use these rules for the generation of texts as well.

This means that, under the rules described by Vorfeld and Mittelfeld, sentence (8), the German translation of sentence (2), has 2 valid variations; sentences (9) and (10).

(8) Lewandowski schoss das Tor in der 65. Minute.
(9) In der 65. Minute schoss Lewandowski das Tor.
(10) Das Tor schoss Lewandowski in der 65. Minute.

Up to this point the argumentation has been exclusively syntactical. This might lead to the conclusion that sentences (8), (9), and (10) are equivalent. However, the fact that things are a bit more complex becomes obvious when semantics or, more precisely, information structure is taken in consideration. In the cognitive process of language production, concepts are annotated by accessibility markers, as we already mentioned when discussing variations of noun phrases. In the production process, accessibility markers also represent additional activation. This means that a concept with a prominent accessibility marker will most probably activate its lexical items faster. The corresponding phrase therefore has a better chance to appear at the beginning of the sentence to be generated. From a formal linguistic point of view, this is the meaning of "information structure", expressed by the formal concepts "Theme-Rheme" and "Focus" [13]. The terms theme and rheme define a sentence by separating known information (theme) and new information (rheme), where the theme would normally precede the rheme. Following this concept, variant (8) would be chosen, if the information that a goal has been scored is already known. The focus is a way to stress the important information in a sentence. It mostly coincides with rheme, but this is not necessarily the case. The concept only works in combination with the concept of an unmarked sentence, where the structure is being changed to emphasize certain elements. One could argue that variant (8) is such an unmarked sentence, because it follows the order of subject-predicate-object. Variant (10) differs from that order and, by doing so, pushes "das Tor" ("the goal") into the focus. This variant could be used as contrast to another action of Lewandowski, e.g. a foul. Currently, we are working on automatically determining the best choice from the available set of sentences.

## 6.6    Recommendations

Automatic generation of texts is worth considering if the purpose the text is given and simple. It is best employed to present data that is available in a structured form, e.g. in a table. Automatic generation of texts is profitable if such a presentation of data is in demand and needs to be repeated regularly.

In order to generate the texts, it is sufficient to make use of templates. Smart variations are fine and necessary but high literacy is not needed and beyond the scope of automatic generation.

In order to increase the quality of texts, human language production strategies can and should be exploited. This includes linguistic means specific for the target language and the use of (simple) ontologies for the representation of knowledge, see also Hoppe and Tolksdorf (Chap. 2).

## 6.7    Summary

In recent years natural language generation has become an important branch in IT. The technology is mature and applications are in use worldwide in many sectors. It is part of the digitalization and automation process which in traditional manufacturing is denoted Industry 4.0. The focus of this article so far was to show what can be expected in terms of generating more sophisticated texts. We advocated for the use of semantics to do so. In combination with an ontology as the knowledge base, the integration of reasoners will allow the derivation of automatically inferred information into the text generation process. The concept of property chains is essential for making this kind of retrieval fast enough. We also have shown how information structure can be used to vary the lexical content of phrases and to find the variation of a sentence that best captures the flow of information and thus contributes to enhancing the quality of generated texts in terms of cohesion and coherence.

## 6.8    The Next Trend: Hyper-Personalization of News

The upcoming trend in media industry is hyper-personalization. To date, most of the news articles are written for a broad audience. The individual reader has to search and select the news that is relevant to her/him. Though many apps already provide news streams for specific domains such as weather, sports or events, none of them create a personalized news stream. In the Google funded project 3dna.news, a novel approach has since been offered as a service in multiple languages. A user is immediately informed by e-mail or WhatsApp if, for example, a specific share she/he is interested in exceeds a given threshold, or when the next soccer game of her/his favorite team begins. In the latter example, she/he is also informed about relevant related information such as the weather conditions expected during the game and about all traffic jams on the way from their home to the stadium.

With hyper-personalization, publishing companies and news portals will be able to provide their readers with new service offerings resulting in a higher customer retention. The news consumer can tailor a subscription to their personal demands and gets the relevant information promptly. Hyper-personalization will also create a new opportunity for in-car entertainment. Currently, radio stations produce one program for all of their listeners. In the future it will be possible to stream the news individually to each car. Generated news will run through a text-to-speech converter and be presented to the driver as an individual radio program. This would also be applicable to in-home entertainment. Amazon's Alexa will allow the user to interact with text generation systems to respond to demands such as: "Alexa, give me a summarized report on the development of my shares!" or "Alexa, keep me informed about the important events of the soccer game of my favorite team!".

However, hyper-personalization potentially increases the danger of "echo chambers" disrupting societies. In addition to this, the resources needed to offer such services are tremendous. The number of news articles that have to be generated is on a much larger scale compared to general news for a broad audience. Also, the news generation process has to run continuously because events triggering the production of a new text could happen any time.

## References

1. Bense H (2014) The unique predication of knowledge elements and their visualization and factorization in ontology engineering. In: Garbacz P, Kutz O (eds) Formal ontology in information systems, proceedings of the eighth international conference (FOIS 2014), Rio de Janeiro, Brazil, Sept. 22–25, 2014. IOS Press, Amsterdam, pp 241–250
2. Bense H, Schade U (2015) Ontologien als Schlüsseltechnologie für die automatische Erzeugung natürlichsprachlicher Texte. In: Humm B, Reibold A, Ege B (eds) Corporate Semantic Web. Springer, Berlin
3. Bock JK (1982) Toward a cognitive psychology of syntax: information processing contributions to sentence formulation. Psychol Rev 89:1–47
4. Bühler K (1934) Sprachtheorie: Die Darstellungsfunktion der Sprache. G. Fischer Verlag, Jena
5. Dell GS (1986) A spreading-activation theory of retrieval in sentence production. Psychol Rev 93:283–321
6. Fromkin V (1971) The non-anomalous nature of anomalous utterances. Language 47:27–52
7. Garrett M (1975) The analysis of sentence production. In: Bower G (ed) Psychology of learning and motivation, vol 9. Academic, New York, pp 133–177
8. Kempen G, Hoenkamp E (1987) An incremental procedural grammar for sentence formulation. Cogn Sci 11:201–258
9. Kempen G, Huijbers P (1983) The lexicalization process in sentence production and naming: indirect election of words. Cognition 14:185–209
10. Levelt WJ (1983) Monitoring and self-repair in speech. Cognition 14:41–104
11. Levelt WJ (1989) Speaking – from intention to articulation. MIT Press, Cambridge, MA
12. Levelt WJ, Roelofs A, Meyer AS (1999) A theory of lexical access in speech production. Behav Brain Sci 22:1–75
13. Musan R (2010) Informationsstruktur. Universitätsverlag Winter, Heidelberg
14. Nübling D, Fahlbusch F, Heuser R (2015) Namen: Eine Einführung in die Onomastik (2 Ausg.). Narr, Tübingen
15. Schriefers H, Meyer AS, Levelt WJ (1990) Exploring the time course of lexical access in language production: picture-word interference studies. J Mem Lang 29:86–102
16. Wöllstein A (2014) Topologisches Satzmodell (2. Ausg.). Universitätsverlag Winter, Heidelberg

# The Role of Ontologies in Sentiment Analysis

# 7

Melanie Siegel

**Key Statements**

1. Sentiment analysis depends heavily on words: sentiment words, negations, amplifiers, and words for the product or its aspects.
2. Words are very important, independent of the chosen analysis method – machine learning or knowledge-based.
3. Words in the context of sentiment analysis are often represented in ontologies.
4. If the sentiment analysis is to achieve more than just classifying a sentence as positive or negative, and if it is needs to identify the liked or hated attributes of a product and the scope of negation, it needs linguistic and ontological knowledge.
5. Ontologies can be used to gain a list of sentiment words.
6. German language has some peculiarities (such as free word order) that makes it impossible to simply apply methods developed for the English language.

## 7.1 Introduction

Who has not read the evaluation of other consumers before booking a trip, buying a book, or cooking a recipe? In recent years, this has become the standard behaviour of consumers. Many consumers also write reviews in sales portals or on Twitter. The consumer thereby has a direct influence on the development of the products. For the companies (such as hotels, authors, producers, etc.), this is a great opportunity to learn more about what is

M. Siegel (✉)
Hochschule Darmstadt, Darmstadt, Germany
e-mail: melanie.siegel@h-da.de

© Springer-Verlag GmbH Germany, part of Springer Nature 2018
T. Hoppe et al. (eds.), *Semantic Applications*,
https://doi.org/10.1007/978-3-662-55433-3_7

important to their customers and what they do not like. Companies can react faster than ever if, for example, something goes wrong, people do not like a new design, a marketing campaign is miscalculated or a product does not work as it should. However, this is only possible if they can quickly extract the information from the opinions expressed by the customers, which requires automatic data processing in the case of large data volumes.

All automatic methods for sentiment analysis work with words: words that express the sentiment ("good"), negations ("not"), amplifiers ("very"), and words that describe the evaluated products or services and their parts and characteristics ("battery", "cleanliness").

This chapter examines the role of the words and phrases (lexical units) in sentiment analysis and the role that ontologies play. Ontologies are understood in this chapter in a broader sense, covering terminologies, thesauri, dictionaries, glossaries etc. We first explain the basics of automatic sentiment analysis, before we look at the lexical units in sentiment analysis. In doing this, we first look at the sentiment words, and, in a second step, at the words designating the aspects and the entities about which opinions are expressed. In both cases, we examine the organization of the lexical units in the sentiment analysis tool, the acquisition of new lexical units, and the adaptation to the domain.

## 7.2 Basics of Automatic Sentiment Analysis

Sentiment analysis is in the field of natural language processing, with information extraction requiring both data mining and text mining. Data mining searches relevant information in data and displays this information in tables or visualizations. Text mining searches relevant information in text data. Text data is so-called unstructured data because it is not recorded in tables or numbers. Information extraction develops methods to transform the extracted information from text data into knowledge. Sentiment analysis is a special type of information extraction. It aims at automatically recognize and classify expressions of opinion (information) in newsgroups, Twitter, Facebook, and forums (language data) and thus extract knowledge about opinions.

The starting point for information extraction is the search for documents answering questions of the following type:

"Where is the email I received last week from Mrs. Mueller?"
"Where can I find information material on my topic?"

These can be resolved by search engines, but the user is not satisfied. There is much more information available that needs deeper investigation. For example, I am interested in whether my research results are being discussed in the research public, and in what way this happens. It is not only relevant how often someone is quoted, but also whether the research results serve as a basis for further research or even serve as a bad example. In this respect, the information in documents is interesting, and it is not enough to just find the documents.

For linguistic research, the question is which methods are best suited for the task. This question is highly interdisciplinary.

In research on information extraction, there are two basic approaches:

The **knowledge-based approach (KB)** uses linguistic methods. Manually generated rules are used for extraction. A disadvantage of this approach is the greater complexity that is connected to the fact that rules are generated manually. However, in a highly modular design, the transferability to new domains and languages is considerably simplified since only modules and not whole processes have to be implemented again.

The approach of **machine learning (ML)** is based on statistical methods in which rules are automatically learned from annotated corpora. Therefore, only linguistic expertise is required. One disadvantage is that annotated training data can be difficult to access. However, if these are available, such a system can be implemented and quickly adapted.

The first task of sentiment analysis is to identify expressions of opinion by distinguishing them from other sentence types, e.g., descriptive sentences. In the KB approach, this is achieved by searching for words and phrases that denote opinions, such as "I believe" or "I think". In the ML approach, there are large sets of sentences for training which have been previously classified manually as containing sentiment or not. From these, the sentiment words and phrases are learned. It should be noted that subjective and emotional expressions are not always expressions of opinion, e.g. "I thought she wouldn't come today" or "I am so sad that I missed the movie!".

Reference [8] classifies the information in expressions of opinion as follows:

1. The entity which is the subject of the opinion expressed, such as "radio".
2. An aspect of this entity that is discussed in the expression, e.g., the reception quality.
3. The opinion, e.g., positive.
4. The person who holds this opinion. This can be relevant if, for example, one wants to find out whether many different persons have expressed their opinion or only some persons have often expressed their opinion.
5. The time at which the opinion has been expressed. This can be relevant, for example, if one wants to observe a change in the customer opinions on a product or the citizen's opinions on a political question.

Based on this classification, the tasks of sentiment analysis are therefore:

1. Identification of the entity
2. Identification of the aspect
3. Identification of the opinion and its polarity, e.g. as positive, neutral, or negative
4. Identification of the person who holds the opinion
5. Identification of the time of expression

First, the sentiment analysis must find the expression that is relevant. In other words, documents or sentences have to be found that deal with the entity that is to be examined. The entity can be expressed in the text in different synonymous variants. Let us suppose that we were talking about a coffee machine named ABC coffee machine. This machine could be referred to as "ABC coffee machine", "ABC Coffee Machine", or "abc coffee machine" in the text. Time and person can often be inferred from metadata. The next step in sentiment analysis is the identification of the opinion and its polarity. The sentiment analysis works with lexical information which was created manually or automatically. For example, the sentiment analysis searches for words and phrases such as "good" or "not bad". These words are first assigned a polarity and then summed in the document or sentence. The most elaborate task is the extraction of the aspect of the entity for which an opinion is expressed. A sentence may contain expressions on several aspects, such as "The battery life of this mobile is great, but it's much too heavy." Aspects can also be implicit, e.g., "too heavy" is the previous statement is inherently referring to the "weight" aspect. To perform aspect-based sentiment analysis, the domain that is being searched must be modeled with its entities and aspects. In addition, it is not enough to work on the document level (e.g., for example, individual Amazon reviews) and, for a precise analysis, one has to go to the sentence level, or even to the phrase level.

## 7.3    Words in Sentiment Analysis

### 7.3.1    Sentiment Words

The basis of the analysis of sentiment and polarity are sentiment words. Often, adjectives such as "bad", "nice", "fast", "robust" express the opinion, but opinions can also be expressed through phrases such as "quickly broken" or "makes me crazy". However, some words are sentiment expressions only in specific contexts. For example, the word "noise", which is neutral in most contexts, has a negative polarity in the context of motors or hotel rooms.

In sentiment analysis systems, sentiment words and phrases may be organized as word lists or ontologies, annotated with their polarity. In most cases, they are organised as word lists. There are general and context-specific sets of sentiment words. Word lists can be created manually and sentiment words can be derived from other lexicons, word nets or from text corpora.

#### 7.3.1.1 Derivation of Sentiment Words from Word Nets

The Opinion Lexicon [5] contains 6800 English positive and negative words. It was set up by manually classifying a number of adjectives from the (manually created) Princeton WordNet [4], and then giving synonyms the same polarity classification and antonyms the opposing classification.

In Ref. [1], the authors present SentiWordNet for the English language, in which sentiment and polarity are annotated. The sets of synonyms (synsets) of Princeton WordNet have been annotated with numeric values for positive, negative, and neutral. Therefore, as in the Opinion Lexicon, a few synsets were annotated manually as positive, negative, and neutral, and then relations such as antonymy were used to generate additional polarities automatically: the antonym of a positive adjective is, for example, negative. The Princeton WordNet contains definitions for each synset. Since these definitions, in turn, contain words that are part of WordNet, Baccianella, Esuli and Sebastiani [1] transfer the polarity to these words.

WordNet, as a general resource containing ontological relations, is used to gain more sentiment words. These sentiment classifications are then returned to WordNet, so that the resource is enriched.

The advantage of this approach is that large sentiment lexicons can be set up quickly and automatically, while the effort of manual "cleaning" is manageable. One disadvantage of this approach is that the classified words are general domain. Therefore, "quiet" in the context of motors is a positive adjective, while it is negative in the context of speakers.

### 7.3.1.2 Derivation of Sentiment Words from Corpora

The goal of gaining sentiment words from text corpora is to take the context dependency of the meaning of many words into account. A training procedure, once implemented, can be easily transferred to other data. In this way, further sentiment words can be obtained.

If sufficient manually annotated text corpora are present, then sentiment words can be learned from them by calculating the probability with which a word occurs in the positive or negative context. Reference [10] describes this method.

This approach is also possible, for example, with Amazon reviews which are displayed with an annotation of 1–5 stars provided by the customers, a procedure which [11] applies to the German language.

The method of semi-supervised learning needs less annotated text data. The basis here is formed by some annotated examples, a large amount of non-annotated text data, and some "patterns". With this method, for example, adjectives coordinated with positive adjectives using "and" are also classified as positive, e.g., "useful and good". Adjectives coordinated with "but" are classified with opposite polarity, e.g., "impractical, but beautiful". Reference [7] describes this process.

Reference [12] renounces the use of text corpora and instead uses a general search engine. Words that often appear in the same phrase with "excellent" and "poor" are searched for and added to the sentiment lexicon.

Reference [6] uses the layout structures of websites with "Pros and Cons" to gain sentiment words.

Reference [3] takes a set of annotated documents from a domain, a sentiment lexicon, and a set of documents from another domain as a basis, and then calculates which sentiment words are domain-independent and which are specific to a domain. The advantage of this approach is that the domain dependency of sentiment words is considered. One disadvantage is that the method is based on annotated text data. One problem with the use of annotated data is that it can yield a moderate level of "Inter-Annotator Agreement" [9].

### 7.3.2  Entities and Aspects

In many cases, it is not sufficient to automatically determine whether a document (a forum or blog post) is positive, negative, neutral, or unrelated. It is also important to find out what exactly is good or bad. Often, one sentence contains several expressions concerning different aspects of an entity, as in this review of a textbook:

> *However, the graphics and tables are very good and certainly serve to better understand the subject.*

A sentiment expression always has a target aspect. This aspect can explicitly appear in the text, like in the example above, but it also can implicitly appear in the text, e.g., *"This car is very expensive"*. This example is a statement concerning the aspect of *price* of the entity *car.*

To capture entities and aspects of a domain, synonyms, meronyms, and hyponyms of the domain must be analysed. Synonyms such as "voice quality" – "speech quality" must be attributed to the same aspect. Synonyms can be domain-dependent. There can also be spelling errors, especially if social media data is analysed.

Meronyms are potentially aspects, that is, the properties and parts of the entity. Hyponyms are subtypes of entities or aspects.

These relations can be captured in an ontology. Therefore, unlike sentiment words, words for aspects and entities are more often organized in ontologies rather than word lists. For a synonym group, a preferred naming can be determined so that all expressions of the synonym group are mapped to the same aspect.

An existing ontology – such as WordNet – can be used to extend the aspect ontology by integrating further synonyms, meronyms, and hyponyms. Another possibility is to exploit texts from the domain. Explicit aspects can be derived from these texts by extracting frequently occurring nouns, which are less common in the general language. The results can be more precise if, for this extraction, only the nouns in sentiment sentences are considered. That is, sentences containing sentiment first have to be identified. In German, nominal compounds are an important source of technical terms, whereas in English multi-word expressions occur more frequently. It is also possible to work with patterns. An example is the pattern of an adjective and a noun, where the adjective is uppercase. This is the case in German if the adjective noun construction is a technical term.

In order to mine aspects of entities (i.e. meronyms), patterns such as "car has", "the car's" and "car is delivered with" can be used. Searching these patterns with a search engine can lead to good results.

When entities and aspects are captured and classified, the analysis of a sentence can be interpreted so that the aspect word closest to the sentiment word is interpreted to be the target aspect of the sentiment. This method, however, often leads to erroneous results, especially when applied to German text. Another approach is the use of a dependency parser that analyses the arguments of expressions, for example, adjectives [2]. A difficult problem is the recognition of co-references in the case of pronouns. Another problem is the detection of implicit aspects, which are usually associated with adjectives, such as "expensive" with price and "beautiful" with appearance.

## 7.4    Recommendations

1. All methods of sentiment analysis need words as their basic information source. Sentiment words and aspect words can be organized into ontologies. This has the advantage that ontological relations can be used for sentiment retrieval and interpretation.
2. Existing ontologies can be used to extend the dictionary of sentiment words.
3. For aspect identification in sentiment analysis, it is necessary to set up or derive an ontology that represents the domain.

## 7.5    Conclusions

Essential core components of sentiment analysis are words and phrases. These are, on the one hand, sentiment words, i.e., expressions of opinions, and, on the other hand, words for aspects and entities in the domain. Sentiment words are usually organized in word lists. In addition, it is common practice to take sentiment words from existing ontologies (such as WordNet). For the domain-dependent sentiment words, methods are used to extract them from annotated text corpora. On the contrary, aspect words are usually organized in ontologies, since ontological relations play an important role. As for the acquisition of aspect words, both general semantic networks and text-based methods are used, because of the strong domain dependency.

## References

1. Baccianella S, Esuli A, Sebastiani F (2010) SentiWordNet 3.0: an enhanced lexical resource for sentiment analysis and opinion mining. In: Proceedings of LREC, Valetta, Malta, pp 2200–2204
2. Chen D, Manning C (2014) A fast and accurate dependency parser using neural networks. In: Proceedings of the 2014 conference on empirical methods in natural language processing (EMNLP), Doha, Qatar, pp 740–750
3. Du W, Tan S, Cheng X, Yun X (2010) Adapting information bottleneck method for automatic construction of domain-oriented sentiment lexicon. In: Proceedings of the third ACM international conference on web search and data mining. ACM, New York, pp 111–120
4. Fellbaum C (1998) WordNet: an electronic lexical. MIT Press, Cambridge
5. Hu M, Liu B (2004) Mining and summarizing customer reviews. In: Proceedings of the ACM SIGKDD international conference on knowledge discovery and data mining (KDD-2004), Seattle

6. Kaji N, Kitsuregawa M (2007) Building lexicon for sentiment analysis from massive collection of HTML documents. In: Proceedings of the 2007 joint conference on empirical methods in natural language processing and computational natural language learning, Prague, Czech Republic
7. Kanayama H, Nasukawa T (2006) Fully automatic lexicon expansion for domain-oriented sentiment analysis. In: Proceedings of the 2006 conference on empirical methods in natural language processing. Association for Computational Linguistics, Sydney, Australia, pp 355–363
8. Liu B (2012) Sentiment analysis and opinion mining. Morgan & Claypool Publishers, San Rafael, California, USA
9. Nowak S, Rüger S (2010) How reliable are annotations via crowdsourcing: a study about inter-annotator agreement for multi-label image annotation. In: Proceedings of the international conference on multimedia information retrieval, Philadelphia, Pennsylvania, USA pp 557–566
10. Potts C (2010) On the negativity of negation. In: Semantics and linguistic theory. Linguistic Society of America, Cornell University, pp 636–659
11. Rill S et al (2012) A phrase-based opinion list for the German language. In: Proceedings of KONVENS 2012, Vienna, Austria, pp 305–313
12. Turney PD (2002) Thumbs up or thumbs down?: semantic orientation applied to unsupervised classification of reviews. In: Proceedings of the 40th annual meeting on association for computational linguistics. Association for Computational Linguistics, Philadelphia, Pennsylvania, pp 417–424

# Building Concise Text Corpora from Web Contents

8

Wolfram Bartussek

**Key Statements**

1. Define your subject matter by ontologies in the sense of information and computer science.
2. Use such ontologies to control which information will be part of your text corpus representing the knowledge you are interested in.
3. Make your text corpora comprehensive yet minimal, i.e., make them concise.
4. Concise corpora can be exploited even with cheap equipment, making them feasible even for very small enterprises.
5. Minimize your own effort by leaving the bulk of maintenance and learning to a semi-supervised learning system such as the CorpusBuilder described here.

## 8.1 Motivation

Taking into account that we are striving for a solution to product surveillance for Small and Medium-sized Enterprises (SMEs), we refrained from attempts to involve any technologies requiring large data centers. Many small enterprises would not be able to afford such services. Additionally, privacy and data security is a big point for SMEs in the medical device sector. They are mainly research driven and work with a lot of sensitive data kept in their quality assurance systems. Thus, they prefer to keep their data on site. Therefore, the technologies presented here work independently of such considerations. They can be run on cheap hardware but they can also be scaled up as needed.

W. Bartussek (✉)
OntoPort UG, Sulzbach, Germany
e-mail: w.bartussek@ontoport.de

© Springer-Verlag GmbH Germany, part of Springer Nature 2018
T. Hoppe et al. (eds.), *Semantic Applications*,
https://doi.org/10.1007/978-3-662-55433-3_8

**Fig. 8.1** Search solution

In the general case, one is not solely interested in data found in the web. Instead, for a complete surveillance solution, one would integrate their own and proprietary data with data being publicly available. Although this integration aspect was one of the predominant aspects of this project, and although it has been thoroughly pursued, it has been intentionally excluded from this chapter.

Here we focus on content that can be retrieved from the internet. Hence, we had to pay attention to which publications to incorporate in the corpora and which to reject in approaching a minimal corpus. Having a minimal corpus would fit the requirement to be able to employ low cost equipment.

But at the same time, nothing should be missing! Engineers in the medical device sector (and elsewhere) are always interested in observed or suspected deficiencies of their products. With respect to medical devices, engineers are very much interested in reported complications and adverse events. It would be more than a glitch if a system for post market surveillance would not be able to find such information securely and timely. Therefore, the corpus is required to be comprehensive and up to date. Technically, this means that we need to have a component that is able to "inspect" the internet to find the right sources and to find them both dependably and quickly.

To meet such requirements, one would traditionally install a web crawler[1], give it the right list of internet addresses (URLs[2]) and let it do its work. Or, if you want to be more specific and the pages are known in advance, you would employ some programmers, and pass them the corresponding URLs. You would then let them write a program using a scraper[3] which extracts the wanted portions of the web pages to be accessed and then pass the data to a search component.

A simplistic overview of the operations of a search solution is given in Fig. 8.1:

Reading the diagram, there is a tool for data acquisition that eventually composes some kind of corpus. Next we have some mechanism to produce an inverted index[4] based on the corpus, called "indexing". The resulting index will be used to extract and interpret data according to some query coming from the end user and finally there is a component presenting the results via some user interface (UI) to the end user.

This picture is an oversimplification in several ways.

---

[1] Web Crawler: https://en.wikipedia.org/wiki/Web_crawler.

[2] Unified Resource Locator: https://en.wikipedia.org/wiki/URL.

[3] Web Scraping: https://en.wikipedia.org/wiki/Web_scraping.

[4] Inverted Index: https://en.wikipedia.org/wiki/Inverted_index.

First, in vertical search solutions[5], reading from the internet is just one way of acquiring relevant data. Therefore, we have to generalize this approach to include any relevant data source. This may be retrievable via internet technologies, but it may however be data that is stored in proprietary customer databases or file systems.

Second, it is misleading to assume that indexing immediately follows data acquisition. In fact most crawlers strip off any markup[6] such as HTML[7] or PDF[8] and store the pure text, although it is clear that markup may carry important semantic content.

Third, indexing is not just about computing a plain inverted index. Instead there are lots of decisions to be taken, e.g., how the index will be structured, which ranking algorithms will be applied up front, which words should be ignored for which languages, which natural languages to support, and so on.

Extraction and interpretation are not further considered here.

Here we focus on data acquisition and forming corpora satisfying the customer needs.

Crawlers have to be told which sites and pages to access. This is mostly done by subject matter experts with a background in computer linguistics or technical writing. They provide an initial seed list, which is normally a plain list of URLs to be crawled.

The crawler will ingest this list, read the reachable pages, and store them as they are or strip off any markup and store the remaining text. Crawlers identify any references to other pages. They read and follow them. There, they also read the referenced pages, thus digging themselves deeper and deeper into the web. There are at least two fundamental problems with such algorithms.

(a)  How can we assure that the pages found are of any value to the customer?
(b)  How can we make sure that the seed list contains everything that is relevant to the customer?

The first question amounts to providing some means to exclude "garbage" or "unwanted" data. This is normally done by specialized filtering algorithms that are programmed ad hoc for special cases. Alternatively, the worthless or offending pages are removed manually. This may be a boring and frustrating endeavor and as such it is highly error prone.

The second question is much harder to answer. There is a widespread belief that universal search engines such as Google, Bing, or Yahoo satisfy the requirement to offer utmost comprehensiveness. However, these search engines can not access all relevant data sources, e.g., internal enterprise data. Even when looking into the internet, we see many open special purpose search engines each with its own query language, partially requiring authentication. It is immediately clear that universal search engines will not take the effort to access such data.

---

[5] Vertical Search: https://en.wikipedia.org/wiki/Vertical_search.

[6] Markup Language: https://en.wikipedia.org/wiki/Markup_language.

[7] Hypertext Markup Language: https://en.wikipedia.org/wiki/HTML.

[8] Portable Document Format: https://en.wikipedia.org/wiki/Portable_Document_Format.

Leaving all these technical considerations aside, there remain questions with respect to knowledge, capabilities and effort of the personnel having to supply all the information needed by such approaches:

(1) Where do these URLs come from?
(2) Who is going to provide them?
(3) Who keeps them up to date?
(4) How would we ensure that we have all relevant URLs?
(5) Who would decide which pages are relevant and which are just garbage?
(6) Given this tremendous effort, would it all be worthwhile?

## 8.2    Objectives

When trying to build minimal, yet comprehensive text corpora from web contents there are obviously several more technical problems to solve. Yet, at the same time, we have to combine the technical solution with a set of standardized processes (in the medical context "standard operating procedures"[9]) involving subject matter experts. These processes are addressed by questions (1) to (5) above. Given that we succeeded in establishing such processes and developing a technical solution doing the bulk work, somebody overseeing the effort as a whole will be confronted with question (6).

Each of the major steps of these processes needs its own quality assurance, obeying some superordinate quality management policy. Here we'll emphasize on the data acquisition step in a currently nonstandard way: the goal would be to automatically find and accept only data we expect to be of high value to the user, addressing relevance as a selection criterion and minimality as a superordinate economic criterion.

Breaking down this global goal, we can conclude that we need some means to:

(1) Find the data of interest.
(2) Evaluate the data found.
(3) Provide some means to do this in a highly automated way.

The key will be an ontology[10,11] that is not only used to precisely define the relevant terminology from the domain of interest, addressing tasks (1) and (2), but also to control the algorithms performing them, addressing task (3).

Although Fig. 8.2 is a simplification, it illustrates the complex iterative workflow of a corpus or search engine admin.

What he would rather like to have is a situation like this:

Obviously, as Fig. 8.3 indicates, there remain some problems to be solved.

---

[9] Standard Operating Procedure: https://en.wikipedia.org/wiki/Standard_operating_procedure.
[10] Ontology: https://en.wikipedia.org/wiki/Ontology.
[11] Ontology in Computer Science: https://en.wikipedia.org/wiki/Ontology_(information_science).

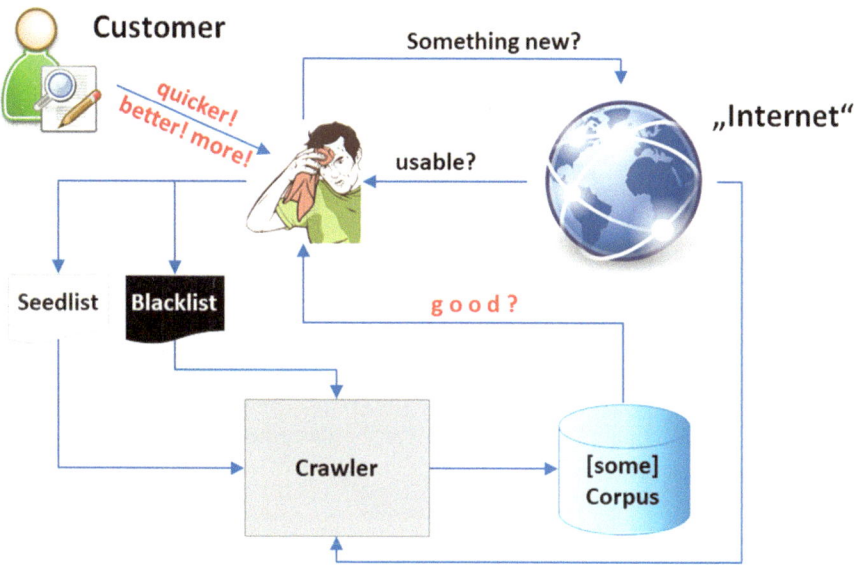

**Fig. 8.2**  Simplified illustration of the workflow of a corpus or search engine admin

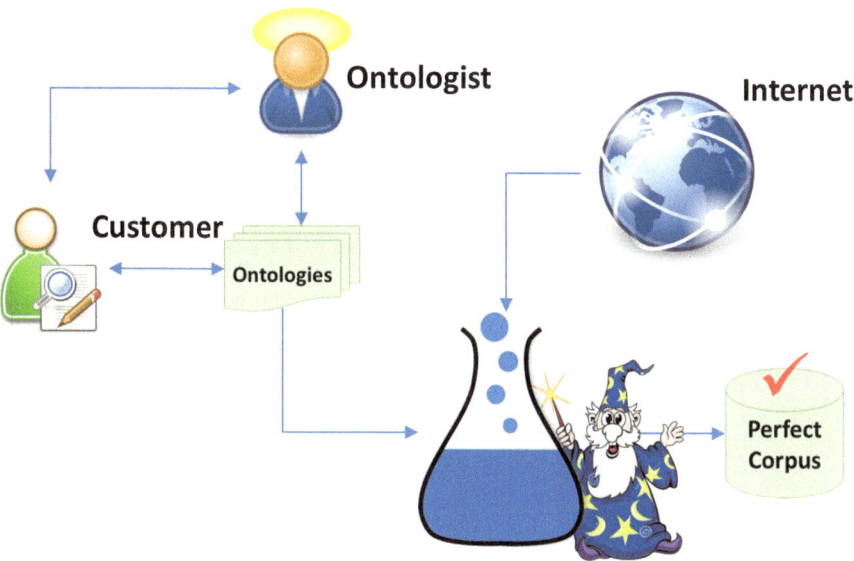

**Fig. 8.3**  Ontologist + Magician = Solution?

## 8.3    Using Ontologies

When building a vertical search engine, we assume that there is some, more or less, well defined set of data available that are either directly fed into an indexer and/or used to build a corpus. Such data can be exploited by tagging[12] relevant terms, indexing them and finally retrieving them and their surrounding documents using one of the known query languages[13]. There are some tacit assumptions in this approach producing substantial effort related to keeping the corpus comprehensive and up to date. These assumptions include:

(1) The set of data comprising the corpus is known in advance.
(2) The vocabulary and terminology used in the corpus is known in advance.
(3) The set of data comprising the corpus can be described simply by listing the data sources to be included.

Practical experience with search engines based on these assumptions suggests however that these assumptions are neither true nor harmless. They are not harmless because they generate considerable effort keeping the underlying data set and the relevant vocabulary and terminology up to date. In many cases, vertical search engines simply die because their value propositions decline by an outdated corpus until there is no economically justifiable perspective in its usage.

Therefore, it is proposed to start out with a means to specify the data set intended to form the corpus instead of explicitly enumerating it and finding a way of specifying how to maintain the underlying vocabulary and terminology in an automated fashion. Such a corpus specification does of course involve a knowledge engineer understanding the proposal and the ultimate goal of the project.

Here we construct such corpus specifications using ontologies in the sense of information and computer science[14,15] [1].

The goal would then be to have a self-regulating continuous improvement spiral (with quality at a certain point in time as its z-axis) ideally warranting an ever optimal specification of the desired corpus.

Ideally one would have a preliminary step to help in getting an initial version of the desired corpus specification. This could be accomplished by feeding a set of characteristic documents to a semantic analysis component resulting in some helpful suggestions for an initial taxonomy. This would then be improved by a knowledge engineer to form the initial specification for the envisaged data corpus.

---

[12] Part of Speech Tagging: https://en.wikipedia.org/wiki/Part-of-speech_tagging.
[13] Query Language: https://en.wikipedia.org/wiki/Query_language.
[14] Ontology: https://en.wikipedia.org/wiki/Ontology.
[15] Ontology in Computer Science: https://en.wikipedia.org/wiki/Ontology_(information_science).

Following any change in such a specification, a concrete enumeration of the data sources is generated. This enumeration is then read by a crawler (or any other means employed for data acquisition) to fetch the desired data and to include them in the current corpus.

## 8.4    The CorpusBuilder

The CorpusBuilder automates many of the steps mentioned above. It is responsible for any kind of data acquisition. It is connected to the internet and possibly various other data sources such as internal databases, intranet chat rooms, and private data, if this helps in augmenting the gathered data. The CorpusBuilder reads the data from various connectors which may need to be specifically customised depending on the data source. It knows these connectors and reads the ontology to derive what to look for.

As outlined above, it is the task of the CorpusBuilder to produce a comprehensive yet minimal corpus of documents to be used by an indexer (such as Apache Lucene[16]). The indexer is needed to provide search functionality for components such as apache Solr[17] or Elasticsearch. For this chapter, the indexer is out of scope.

To minimize the effort in setting up such machinery and in maintaining it, CorpusBuilder is controlled by an ontology that ultimately defines what to be included in the corpus (comprehensiveness) and what should not be part of it (minimality). Therefore, the ontology controls the behavior of the CorpusBuilder.

The most important component of the CorpusBuilder is the Prospector. It is the only component connected to the Internet knowing HTML and how to search the Internet in various ways. The components contained in the red rectangles of Fig. 8.4 comprise the Prospector. Important other parts like the inclusion of private or proprietary data are not considered here since they do not provide web content and require specific customisation.

## 8.5    The Prospector Component

The Prospector is the core component of the CorpusBuilder. The Prospector uses the Internet as its sole data source. It has been developed to meet the requirement "to find the right sources and to find them both dependably and quickly". The Prospector is controlled by an ontology, defining what to look for. In the early stages of its development, the ontology was developed by a subject matter expert (e.g., a medical device engineer) and an expert ontologist, having the specific mathematical background and the knowledge of the needed formalisms and tools. Currently, we build a referential corpus in advance by exploiting a known set of about 400 URLs identifying documents that have been

---

[16] Apache Lucene Core: https://lucene.apache.org/core/.

[17] Apache Solr: http://lucene.apache.org/solr/features.html.

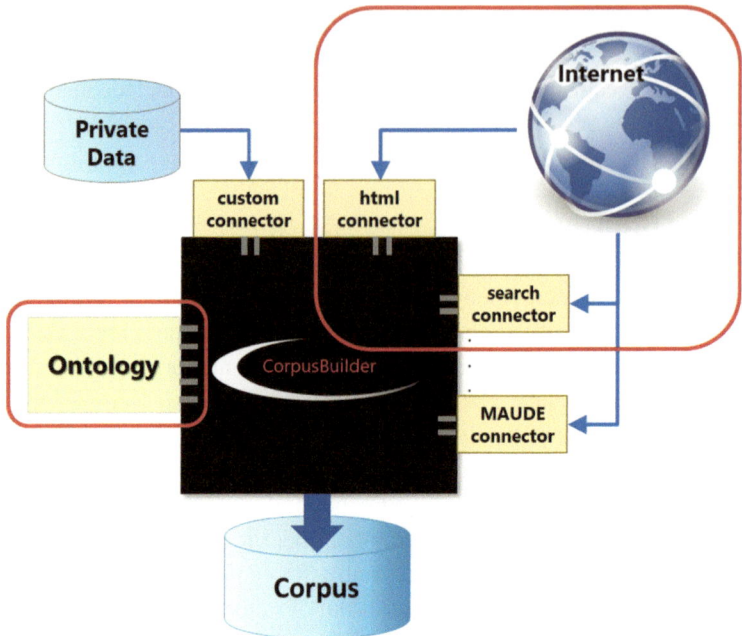

**Fig. 8.4** CorpusBuilder

confirmed to be relevant for the subject matter. From these documents we calculate a
model and a vocabulary (sorted by frequency) using the Word2vec package gensim[18].

Independent of this automated analysis, we build a second vocabulary which is based
on just the personal knowledge of subject matter experts. These two vocabularies are then
automatically combined and used to build the needed ontology by connecting the terms
through characteristic abstractions and other relations.

From this amended ontology we generate a huge number of queries that are used by the
Prospector to search the internet via universal search engines and also by publicly available specialized search engines for the domain of interest. Currently this is for medicine in
general or special disciplines within medicine like surgery, enterology and the like.

The Prospector analyses each retrieved page to find out if it offers search functionality
and if it does, it puts the underlying URL on a proposed list of potential additional search
engines which may be used to find even more relevant publications. This list of potential
additional search engines has to be evaluated regularly by subject matter experts.

Confirmed findings of the subject matter experts are communicated to the software
development team. The software development team then checks the query syntax and
either assigns the newly found search engine to a pool of known search engines with identical syntax or they make up a new pool if the effort seems to be worthwhile. All of this

---

[18] gensim, Topic Modeling for Humans, https://radimrehurek.com/gensim/.

**Fig. 8.5** Feedback loop
schematic

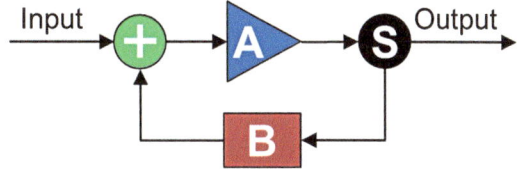

happens in an infinite loop involving the Prospector component, occassionally the subject matter experts and very rarely the software developers.

The Prospector is organized like a control circuit with a feedback loop (See Fig. 8.5).

The "Input" consists of lists of URLs, the seed lists i.e. lists of URLs. Component "A" (including the controller and plant), fetches the contents and extracts further links (URLs). The sensor "S" uses the corpus ontology to check if the content from component "A" should be included in the corpus or not. If the sensor is in doubt, it passes the document to component "B". This backwards oriented component "B" inspects the output in doubt and provides feedback. Component "+" filters out incoming blacklisted URLs and adds new as well as whitelisted URLs.

This process is given in more detail in Fig. 8.6.

The arrows in Fig. 8.6 denote data flows. The storage symbols (cylinders) alongside the arrows denote the data flowing in the direction of the arrow. The symbols "+", "A", "S", and "B" denote the components from the schematic in Fig. 8.5.

The ***Prospector Input Provider*** is a rather sophisticated component because it derives queries from the Prospector Ontology and formats them such that the corresponding search engines will be able and "willing" to process them. This depends, for example, on specific syntaxes, frequency of queries tolerated, and the allowed length of queries, as the length of such queries may be several thousand characters.

Queries may not be accepted at any time and at any frequency, and valid queries may be rejected (no "willingness" to process them). Thus the Prospector Input Provider has a time management component to satisfy timing requirements.

Several search engines restrict the length of accepted queries. Consequently the Prospector Input Provider has a component that is able to split long queries according to length requirements. The shortened queries are then transferred to the time management component.

Some universal search engines are free if they are not used too frequently by automated systems. Frequency considerations also go into the time management component.

Especially in the medical sector there are a lot of special purpose search engines with really good contents. However, almost all of them have their own query syntax and of course, its specific length restrictions. Thus, there have to be multiple algorithms to generate valid queries for the search engines at hand.

The Prospector Ontology is rather coarsely grained to capture also content that may be new to the subject matter or not yet known to the subject matter experts operating the CorpusBuilder.

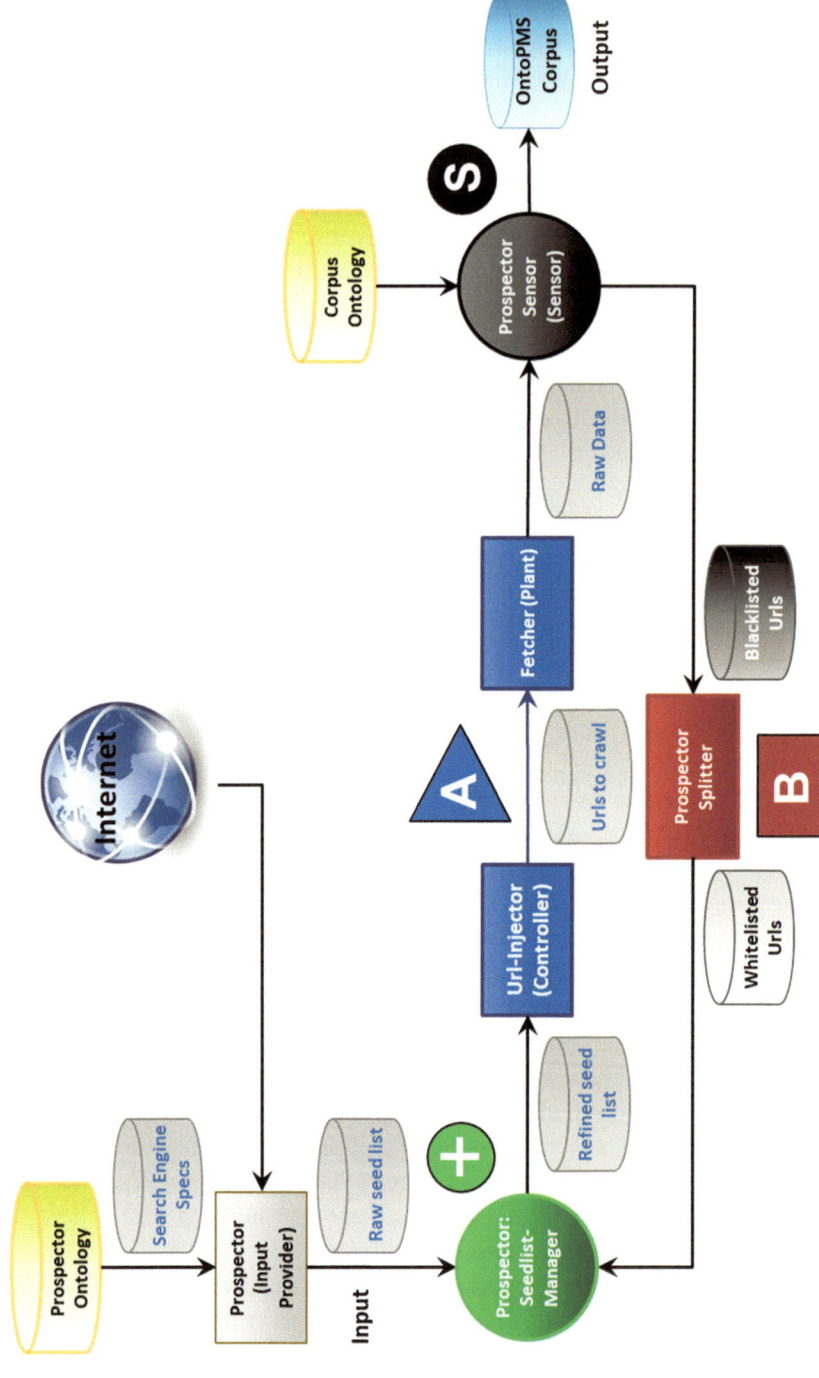

**Fig. 8.6** The prospector feedback loop in detail

The **Prospector Seedlist Manager** takes the URLs of the Prospector Input Provider and combines them with the whitelisted URLs from feedback. It also removes all black-listed URLs identified by the feedback component "B".

Component "A", with its **URL Injector** (the Controller) and **Fetcher** (the Plant), is basically implemented by a standard crawler. It is controlled by the seed list that has been cleansed by the Prospector Seedlist Manager employing the knowledge available so far.

The **Prospector Sensor** is a filter strictly following the rules imposed by the Corpus Ontology. It uses the Corpus Ontology to single out all documents that we want to become part of the corpus. All URLs underlying the documents are marked as being hits, going into the corpus, or failures to be excluded in future.

This information is used by the feedback component "B", the **Prospector Splitter**. This component reduces URLs to domain names[19]. If a URL is marked as a failure its domain will be further investigated. If there are not any relevant documents, the domain of the URL will be blacklisted. Conversely if the domain of a URL marked as a hit returns further hits, it is whitelisted. The results are fed back to the Prospector Seedlist Manager and the loop continues.

So far, we looked only at the automated parts of the system. There are however two parts that have to be provided manually: the two aforementioned ontologies. These ontologies have to be maintained by subject matter experts, the maintainers, meaning that the CorpusBuilder is a semi-supervised learning system.

## 8.6    Alerts

There is one more rather small but quite important component residing in the Prospector Sensor: the **Alerter**. Every time the sensor identifies a hit to be reported to the maintainers of the PMS system, it issues an alert and transmits it via email. The following is a real life example:

---

**Your search pattern:** "OTSC Fistula"~400

  **Title:** Over-the-scope clip closure for treatment of post-pancreaticogastrostomy pancreatic fistula: A case series

  **Snippet:**

  … of patients with post-pancreaticogastrostomy pancreatic fistula in whom OTSC were used as endoscopic treatment … Over-the-scope clip closure for treatment of post-pancreaticogast

  **Language**: en **Date**: 2017-05-26T11:38:21.538Z

  **Url**:    https://insights.ovid.com/digestive-endoscopy/digend/9000/00/000/scope-clip-closure-treatment-post/99670/00013150

---

[19] Domain Names: https://en.wikipedia.org/wiki/Domain_name.

The "search pattern" is one of the numerous queries generated from the corpus ontology. The query is executed on the indexed corpus. The alerter component extracts the title, a snippet where the query applies, the Language, Date and URL. Maintainers use this information to check whether the CorpusBuilder is "on track" in an acceptable way. If it deviates from the intended purpose, the ontology or the blacklist have to be adjusted. Thus the alerter plays a vital role in keeping the system as a whole oriented towards its intended purpose and up to date.

## 8.7    Technology

All components of the CorpusBuilder that were not programmed in the course of the project are open source components. Currently there are several programming languages in use:

- PHP has been used to program the Prospector component as far as search engine and URL-management is involved.
- Python 3 has been used to program the linguistic analyses within the CorpusBuilder.
- Java has been used by IMISE (University of Leipzig) to program query generation from ontologies [2] and by IntraFind for their Elasticsearch-plugin.
- TypeScript and, as part of packages, JavaScript is used for the user interface

There are numerous additional open source packages in use, partly because the major packages depend on them. The major packages are:

- Apache Nutch[20] is used as the crawler.
- Elasticsearch with Lucene to index the documents identified as relevant.
- NLTK[21] and spaCy[22] are used for linguistic analyses.
- MySQL is used as data management system for URL and search engine data.
- Angular 2[23] is used for programming the web user interface (which is not part of the CorpusBuilder).

All of the components of the CorpusBuilder are implemented and run as server components with a RESTful interface to be used by web interface frameworks such as Angular 2, which is used as the preferred UI framework throughout the project.

For ontology editing we used Protégé (https://protege.stanford.edu/).

---

[20] Apache Nutch Web Crawler: http://nutch.apache.org/.

[21] NLTK Natural Language Toolkit: http://www.nltk.org/.

[22] Industrial-Strength Natural Language Processing: https://spacy.io/.

[23] Angular cross platform web framework: https://angular.io/.

## 8.8    Current State and Future Work

Currently ontologies are developed by subject matter experts and specialized ontologists using their current body of knowledge. To become aware of new developments in the observed market, the Prospector issues alerts addressed at the subject matter experts. These alerts, which have beforehand passed all checks with respect to relevance, are then investigated by the subject matter experts.

If they detect new relevant terms in truly relevant documents, they will improve the ontologies respectively. If they detect that an alert identified a document that should actually be blacklisted, they will also improve the ontologies and/or update the blacklist directly. Although this procedure has been quite successful in the past we are not completely satisfied.

We hope to improve this process significantly by utilizing linguistic analyses both on the corpus and on individual documents. Vocabularies and the ranking of terms should be used to give the right hints to the subject matter experts where to look, and how to improve their ontologies.

## 8.9    Summary

The new European Medical Device Regulation (MDR [3]) has been put into force in May 2017 and requires in article 4:

> *Key elements of the existing regulatory approach, such as the supervision of notified bodies, conformity assessment procedures, clinical investigations and clinical evaluation, vigilance and market surveillance should be significantly reinforced, whilst provisions ensuring transparency and traceability regarding medical devices should be introduced, to improve health and safety.*

This general requirement is detailed later on in the MDR at various places and in the annexes. Following these requirements we developed a "post market surveillance" (PMS) system for small and medium enterprises in the medical device sector. The system is at its core a multi-language system and can be adapted to the needs of medical device manufacturers in any country. Although it has been developed with the MDR in mind, it can just as well be applied outside the European Union and for completely different application areas.

Currently we have about 500.000 different medical devices on the European market. Each of them satisfies individual needs of the envisaged patients, with individual choice of materials and design features. To broaden the scope of applicability, it is immediately clear that such a system should have a common kernel and a set of components whose behavior depends on rules capturing all aforementioned individual properties of the medical device at hand and its target market. Consequently, to support, e.g., a medical device manufacturer, this set of rules has to be established upfront, evaluated, and finally put in production mode.

With respect to publicly available information the PMS-system will roam the internet for relevant data like scientific publications about use cases of the medical device – including risk related information like complications and adverse events.

As opposed to the traditional approach of manually maintaining an extensive list of URLs, we construct an ontology specifying the subject matter under consideration. This ontology is then used to generate appropriate queries for various search engines. These deliver the URLs we are interested in. So, instead of maintaining URL lists manually, we replace this boring and error prone task by setting up and maintaining an ontology.

If the subject matter changes e.g., due to new and unforeseen developments, we adapt the ontology correspondingly. Thus, we use the ontology to control the behaviour of the tools used to crawl or scrape the Internet.

Although we succeeded in substantially reducing clerical work in manually searching the Internet and maintaining URL lists, one will still have to check whether the results are satisfying, and adjust the ontology and the blacklists to yield optimal results.

After an initial phase of about 3 months for setting up the ontology and closely observing the results, the effort for further maintenance reduces drastically. Maintenance is mainly triggered by alerts issued by the CorpusBuilder. These alerts tell the maintainer what and where new publications have been found, so these alerts are inspected and evaluated to derive maintenance measures.

Since we are targeting SMEs, we put an emphasis on providing technologies that would be affordable for even very small enterprises. Currently the whole PMS system could be run on a modern laptop computer. This would not be possible with vast amounts of data. Therefore we strived to reduce the data volume to its absolute minimum. Currently we gathered about 2.700.000 publicly available documents and we suspect that there is still a certain amount of irrelevant items hidden in this collection. Although this amount is still no challenge for a modern laptop, very recent linguistic analyses seem to indicate that there is further potential to reduce it substantially.

Currently ontologies are the major means to ensure comprehensiveness and the Prospector with its sensor and feedback loop takes care of minimality.

The CorpusBuilder also includes a set of data connectors that support the inclusion of the manufacturer's proprietary data like their own databases and flat files. All these data are fed into a specialized search engine which ultimately provides the desired benefit.

## 8.10   Recommendations

1. Write a rough description of your domain of interest.
2. Gather a fair amount of documents that best embody your domain interest.
3. Use the URLs of these documents as the initial white list (see Fig. 8.6).
4. Derive a fine grained ontology, modeling the selected documents and your domain of interest to control the Prospector Sensor (see Fig. 8.6).

5. From the fine grained ontology derive a coarsely grained ontology to control the Prospector Input Provider (see Fig. 8.6).
6. Run the CorpusBuilder and observe the generated output, blacklist, and whitelist.
7. Adjust your ontologies, the blacklist, and the whitelist if needed.

Steps 4 and 5 may also be executed in reverse order and in practice one will loop through them several times. Similarly, one will loop through steps 6 and 7. In the beginning, this will be repeated within hours, later weekly and in stable operation, every 3 months, for example.

**Acknowledgements**  Acknowledgements go to all participants of the OntoPMS consortium. With respect to ontologies, accompanying work flows, and available technologies I would like to thank Prof. Heinrich Herre, Alexandr Uciteli, and Stephan Kropf from the IMISE at the University of Leipzig for many inspiring conversations. I wouldn't have had much chance to understand medical regulations in Europe without the help of the novineon personnel Timo Weiland (consortium project lead), Prof. Marc O. Schurr, Stefanie Meese, Klaus Gräf, and the quality manager from Ovesco, Matthias Leenen. The participants from the BfArM, the German Federal Institute for Drugs and Medical Devices, with Prof. Wolfgang Lauer and Robin Seidel helped me understand the MAUDE[24] database and how to connect it to the CorpusBuilder. IntraFind (Christoph Goller and Philipp Blohm) developed an ingenious enhancement to the search engine exploiting the corpus; and MT2IT (Prof. Jörg-Uwe Meyer, Michael Witte) will provide the structures of the overall system where the CorpusBuilder will be embedded. I also would like to thank my colleagues at OntoPort, Anatol Reibold and Günter Lutz-Misof for their astute remarks on earlier versions of this chapter.

# References

1. Herre H (2010) General formal ontology (GFO): a foundational ontology for conceptual modelling. In: Poli R, Healy M, Kameas A (eds) Theory and applications of ontology: computer applications. Springer, Dordrecht, pp 297–345
2. Uciteli A, Goller C, Burek P, Siemoleit S, Faria B, Galanzina H, Weiland T, Drechsler-Hake D, Bartussek W, Herre H (2014) Search ontology, a new approach towards semantic search. In: Plödereder E, Grunske L, Schneider E, Ull D (eds) FoRESEE: Future Search Engines 2014–44. Annual meeting of the GI, Stuttgart – GI edition proceedings LNI. Köllen, Bonn, pp 667–672
3. Medical Device Regulation (EU) 2017/745 of the European Parliament and of the Council of 5 April 2017 on medical devices, OJ. L (2017) pp 1–175

---

[24] MAUDE – Manufacturer and User Facility Device Experience: https://www.accessdata.fda.gov/scripts/cdrh/cfdocs/cfmaude/search.cfm.

# Ontology-Based Modelling of Web Content: Example Leipzig Health Atlas

**9**

Alexandr Uciteli, Christoph Beger, Katja Rillich,
Frank A. Meineke, Markus Loeffler, and Heinrich Herre

**Key Statements**

1. The realisation of a complex web portal, including the modelling of content, is a challenging process. The contents describe different interconnected entities that form a complex structure.
2. The entities and their relations have to be systematically analysed, the content has to be specified and integrated into a content management system (CMS).
3. Ontologies provide a suitable solution for modelling and specifying complex entities and their relations. However, the functionality for automated import of ontologies is not available in current content management systems.
4. In order to describe the content of a web portal, we developed an ontology. Based on this ontology, we implemented a pipeline that allows the specification of the portal's content and its import into the CMS Drupal.
5. Our method is generic. It enables the development of web portals with the focus on a suitable representation of structured knowledge (entities, their properties and relations). Furthermore, it makes it possible to represent existing ontologies in such a way that their content can be understood by users without knowledge of ontologies and their semantics.

A. Uciteli (✉) · C. Beger · K. Rillich · F. A. Meineke · M. Loeffler · H. Herre
University of Leipzig, Leipzig, Germany
e-mail: auciteli@imise.uni-leipzig.de; christoph.beger@imise.uni-leipzig.de; katja.rillich@imise.uni-leipzig.de; frank.meineke@imise.uni-leipzig.de; markus.loeffler@imise.uni-leipzig.de; heinrich.herre@imise.uni-leipzig.de

© Springer-Verlag GmbH Germany, part of Springer Nature 2018
T. Hoppe et al. (eds.), *Semantic Applications*,
https://doi.org/10.1007/978-3-662-55433-3_9

## 9.1    Introduction

The field of systems medicine[1] deepens the understanding of physiological and pathological processes in order to derive new diagnostic and therapeutic approaches. In addition to clinical data, extensive genomic data are processed. Data from various studies are also collected, analysed and combined. The methods for analysing and modelling are very closely linked to the data. The scientific gain in knowledge cannot be passed on alone through publications since the publication of methods and data is equally important. The preparation of the data provided by a local research group for a broad user community requires comprehensive research and data management, but also a carefully thought-through data-sharing concept.

The Leipzig Health Atlas (LHA)[2], launched in 2016, delivers a multifunctional, quality-assured and web-based repository of health-relevant data and methods (models and applications) for a broad research population. Partner teams in Leipzig contribute extensive data, methods and experience from clinical and epidemiological studies, research collaborations in systems medicine, bioinformatics and ontological research projects. The LHA brings together ontologists, modellers, clinical and epidemiological study groups, bioinformaticians and medical informaticians.

The LHA manages extensive content and representation metadata on the publications, data and methods of the participating research projects. The web portal of the LHA serves as a shop window and marketplace for data and innovative methods (models and applications). Depending on the legal regulations, clinical and genomic microdata can be downloaded directly or via appropriate access controls. Where applicable, applications and models can be run interactively in the portal and evaluations can be carried out on an ad hoc basis.

The creation of a complex web portal, including modelling the content, is a challenging process. The contents describe different interconnected entities and have a complex structure. The entities and their relations have to be systematically analysed, and the content has to be specified and integrated into a content management system (CMS). The ontology provides a suitable solution for modelling and specifying complex data and their dependencies. However, since automated import of ontologies in web portals is lacking, we have focused on this problem.

In order to describe the metadata on the projects, publications, methods and datasets to be represented in the LHA portal, we developed an ontology. Based on this ontology, we implemented an ETL (extract/transform/load) pipeline (Fig. 9.1) that allows the specification of the portal's content and its import into the CMS Drupal (Version 8).

---

[1] *"Systems Medicine is the implementation of Systems Biology approaches in medical concepts, research and practice. [...]"* (https://www.casym.eu/what-is-systems-medicine/).

[2] Funded by the German Ministry of Education and Research (reference number: 031L0026, program: i:DSem – Integrative Datensemantik in der Systemmedizin).

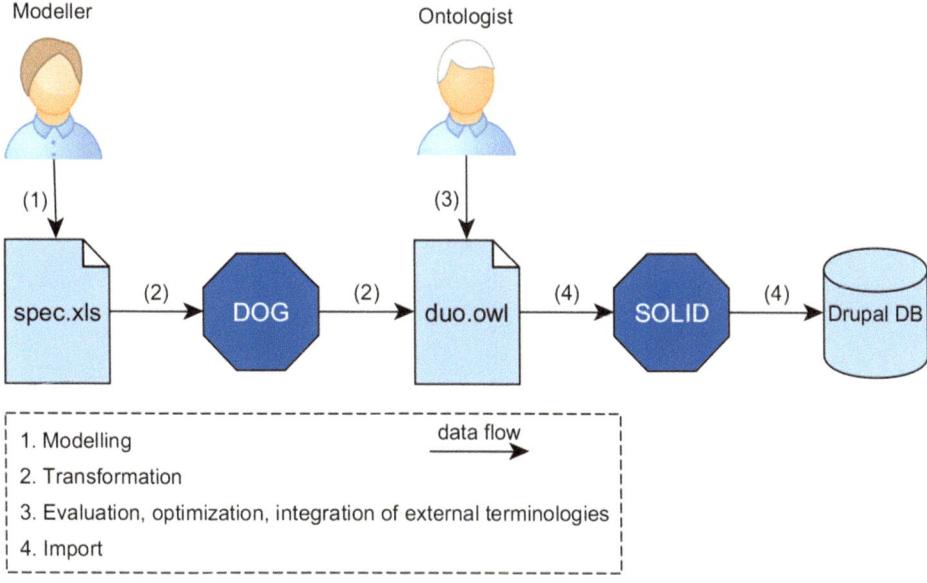

**Fig. 9.1**  Pipeline for the import of content in Drupal

The pipeline consists of the following four steps:

1. Modelling of the content using a spreadsheet template (Fig. 9.3).
2. Transformation of the domain-specific entities from the spreadsheet template to the ontology using the Drupal Ontology Generator (DOG).
3. Optional optimization of the ontology using an ontology editor including the import of external ontologies/terminologies.
4. Importing the ontology into Drupal's own database using the Simple Ontology Loader in Drupal (SOLID).

The approach and the individual components will be discussed in detail in the following sections.

## 9.2    Content Specification of the LHA Portal

For the metadata specification of the projects, publications, data and methods, a metadata model of the LHA was developed (Fig. 9.2). The metadata model consists of three inter-connected levels (entity types).

Publications can be assigned to several projects. The datasets (OMICS-datasets [1], clinical trial data, and other specific datasets) and associated methods are mostly assigned to publications and form the lowest level for capturing the accompanying metadata. It is possible to refer comprehensive datasets to more than one publication. References between entities are realized using IDs.

The collection and processing of metadata is based on a spreadsheet software. This allows for a flexible approach in the development phase.

The metadata model (Fig. 9.2) is implemented in spreadsheets (Fig. 9.3) and it forms the basis for the collection of metadata. The spreadsheet queries specific information in the individual worksheets for the respective entity types (project, publication, OMICS-dataset, clinical dataset, and method).

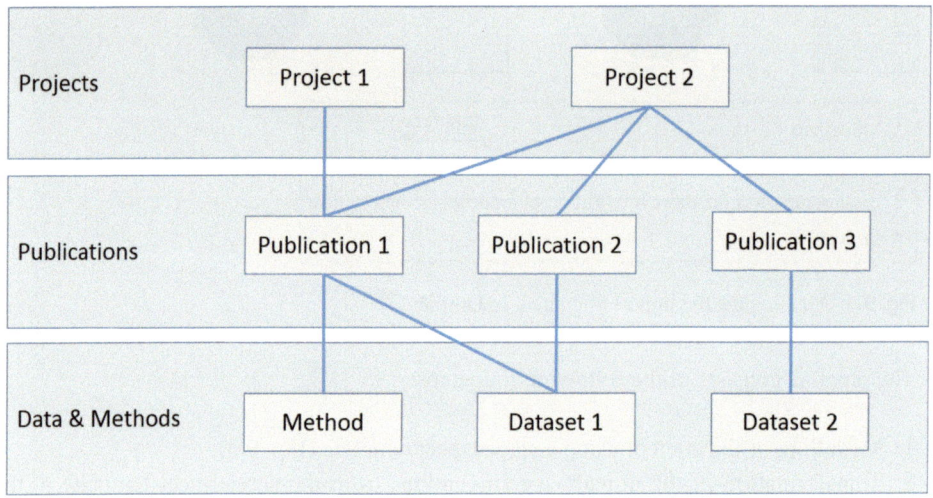

**Fig. 9.2** Metadata model of the LHA

| Merkmal | Format | Frage | Antwort |
|---|---|---|---|
| title | text | Bitte geben Sie den Titel des Projektes/der Studie an: | Leipzig Research Center for Civilization Diseases. Head and Neck Group |
| field_study_shortcut | id | Gibt es ein Projekt-/Studienkürzel? Falls ja, wie heißt das Kürzel? | LIFE-HNG |
| field_study_grouplink | text | Link zur Webseite der Studiengruppe: | http://life.uni-leipzig.de/ |
| field_founding_year | integer | In welchem Jahr wurde das Projekt/die Studie begonnen? | 2012 |
| field_publication_date | date | Publikationsdatum: | 24.06.2015 |
| field_sponsor | text list | Durch welche(n) Förderer wird das Projekt/die Studie finanziell unterstützt? | Leipzig Research Center for Civilization Diseases (LIFE) University Leipzig European Union, the European Fund for Regional Development (EFRE) Free State of Saxony |
| field_disease | taxonomy_reference list | Bitte nennen Sie einige Stichwörter zu diesem Projekt/der Studie (bitte nennen Sie dabei auch die erforschten Krankheiten): | head and neck squamous cell carcinoma\| human papillomavirus |
| field_author | node list | Autoren (werden von uns ergänzt): | Wichmann G\| Rosolowski M\| Krohn K\| Kreuz M\| Boehm A\| Reiche A\| Scharrer U\| Halama D\| Bertolini J\| Bauer U\| Holzinger D\| Pawlita M\| Hess J\| Engel C\| Hasenclever D\| Scholz M\| Ahnert P\| Kirsten H\| Hemprich A\| Wittekind C\| Herbarth O\| Horn F\| Dietz A\| Loeffler M\| Leipzig Head and Neck Group (LHNG). |

Project | Publication | OMICS-Dataset | Clinical Dataset | Method | ⊕

**Fig. 9.3** Example of a spreadsheet for metadata collection

The definition of which meta information must be queried is based on different sources and standards. The higher-level archive functionality of the LHA is based on the OAIS (Open Archival Information System) ISO standard [2]. The OAIS metadata model used for the LHA was supplemented in further steps and compared with broadly generic publication-related standards (for example, schemas of the MEDLINE/ PubMed database [3]). If a publication is listed in MEDLINE, it is sufficient to enter the MEDLINE ID, and the corresponding bibliographical data can be completed automatically. For the definition of domain-specific properties of genetic and clinical data, schemes of existing data portals, e.g., GEO [4], TCGA [5], cBioPortal [6] and CGHUB [7], were taken into account. Properties perceived as missing were added. The resulting metadata list was reviewed and revised based on application to existing projects with a wide range of entity types (e.g., publication, dataset, method) and data types (e.g., text, date, number, reference) jointly with the responsible scientists. In doing so, irrelevant requirements were eliminated, the metadata queries were linguistically defined and new aspects were included. In order to check the data acquisition and display of the metadata, the data tables from each entity type and from each data type were filled with several examples and loaded into the content management system using our pipeline (Fig. 9.1).

In addition to the bibliographic data, the spreadsheet template gather further information on the contents of the projects, publications and data records in such a way that the context of a project or the publications and associated datasets is known before the data is downloaded or a request for access to the data is made. At the project level, the content of this website includes, e.g., links to existing project websites, information on the objectives of the projects, the funding and sponsors, information regarding specific questions on data management and biometrics, as well as annotations with concepts of external terminologies. On the publication level, the abstract, the link to the original publication, the data on sponsors, relevant keywords and authors are recorded. At the dataset level, the content of the datasets including case numbers and design is briefly described. Additionally, information on the responsible scientists (name, address, email address, perspective ORCID (an identifier for scientists)) are collected at all levels.

Depending on the context, the metadata itself is entered as a link, text, text enumeration separated by a concatenation character (vertical stroke), numeric entries or as a date.

## 9.3    Ontological Architecture

We developed the Drupal Upper Ontology (DUO), which models the standard components of Drupal (field, node, file and vocabulary). According to the 3-Ontologies-Method [8], DUO is a task ontology, i.e., an ontology for the problem to be solved by the software. Furthermore, we implemented a domain ontology, the Portal Ontology of LHA (POL), which is embedded in DUO and used to model the content of the portal. For the integration and formal foundation of both the task and domain ontologies, we used the General Formal Ontology (GFO) [9, 10] as a top-level ontology (Fig. 9.4).

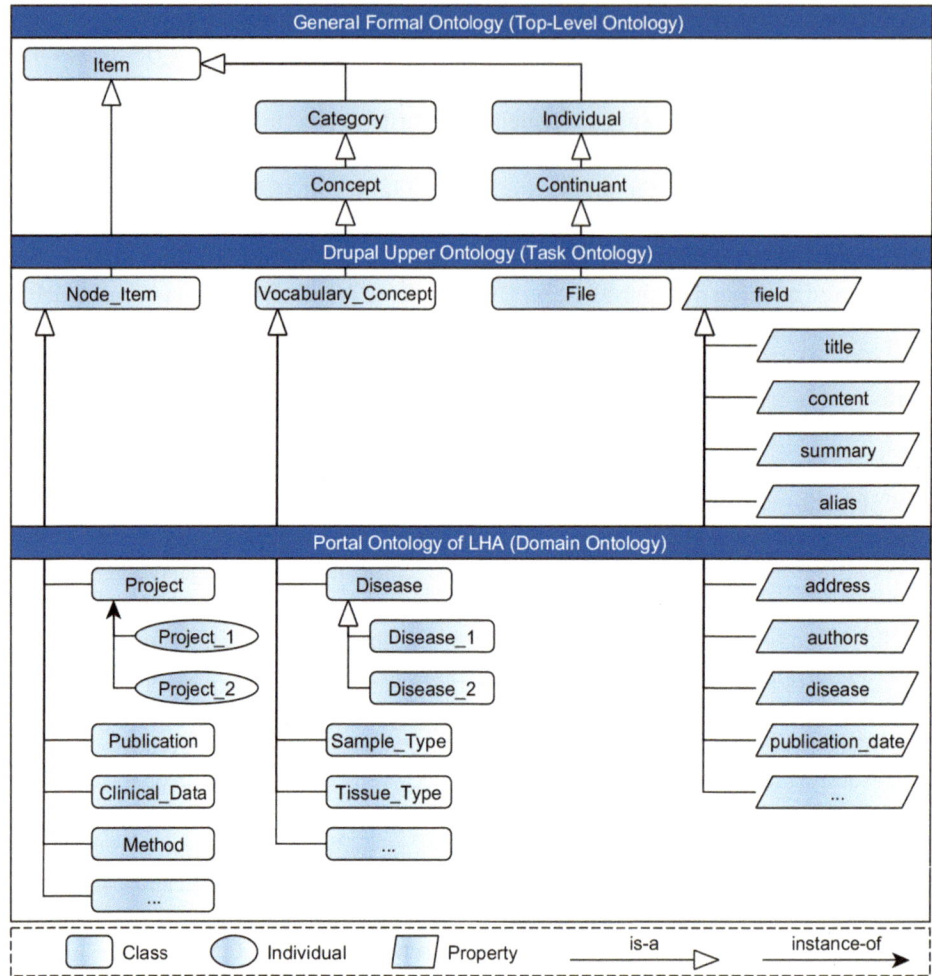

**Fig. 9.4** Ontological architecture

According to GFO, we distinguish between symbolic structures (e.g., the content of web pages, such as text and images) and entities (categories or individuals, such as persons or projects) that are represented by the symbolic structures. For the sake of simplicity, we only model the entities to be represented in the web portal, while their representations on the web pages are generated. To reference entities of a particular ontology the notation <ontology_name>:<entity_name> is used in this chapter. For example, the class Node_ Item from DUO is designated as duo:Node_Item.

Since both individuals and categories can be represented in a portal, we derive the class duo:Node_Item, to model entities to be represented, from the class gfo:Item, which has the classes gfo:Individual and gfo:Category as subclasses. The class duo:Vocabulary_Concept is used for integrating the concepts of external ontologies/terminologies (e.g., disease or phenotype ontologies) and is derived from gfo:Concept. We consider the files (duo:File) as continuants (gfo:Continuant) in the GFO sense, since they are concrete individuals with a certain lifetime.

The categories of DUO are specialized and instantiated in POL. Various entity types, e.g., pol:Project, pol:Method and pol:Clinical_Data are defined as subclasses of duo:Node_Item, and concrete instances of these classes are created and linked. In addition, external terminologies (such as classifications of diseases) are referenced in POL, so that the POL entities can be annotated with their concepts.

Both the Drupal fields (such as "title" or "content") and the user-defined domain-specific fields (such as "address", "author" and "disease") are modelled as annotation properties and used for description and linking of the instances in POL.

## 9.4    Drupal Ontology Generator (DOG)

We developed the Java application Drupal Ontology Generator (DOG) to transformation the domain ontology from the spreadsheet template to Web Ontology Language (OWL).

When reading a completed spreadsheet template (Fig. 9.3), the DOG interprets each data sheet as a representation of one or more instances of a particular type/class. If a datasheet, for example, is called "Project", then specific individual projects are represented/described on this datasheet. For each datasheet name, the DOG generates a subclass of the class duo:Node_Item (if the class does not already exist) and creates the appropriate instances of that class based on their properties. The DOG passes through all specified properties line by line and varies its procedure depending on the defined format.

If "id" is selected for the format field of a property in the spreadsheet, the value of the property is used to generate the Internationalized Resource Identifier (IRI) of the instance and allows referencing of the instance in the same or other files.

When specifying one of the default data types ("text", "integer", "double", "date") in the format column, an annotation is created. The annotation property is used with the name specified in the column "Merkmal" (the German for "feature"), the data type defined in the column "Format", and the value entered in the column "Antwort" (the German for "answer"). If the annotation property with the desired name does not already exist, it is generated as a subproperty of duo:field.

If "taxonomy_reference" or "taxonomy_reference list" is selected in the format column, a subclass of the class duo:Vocabulary_Concept is generated, which is named using the property (without the "field" prefix, e.g., "Disease" from "field_disease") and represents the root node of the corresponding vocabulary. Subclasses of the root class are created for all values of the property (for example, various diseases). Next, an annotation is created that links the corresponding instance to the vocabulary class representing the desired disease (e.g., a link joining a project instance with the vocabulary concept of the disease it deals with). In this way, it is modelled that the current instance is tagged/annotated by certain concepts of defined vocabularies.

The "node", "nodelist", "node_reference", and "node_reference list" formats are used to create links between individual instances. For this purpose, we also use annotation properties. In addition to the defined relations, all instances specified in a spreadsheet file are linked together. The names of the required annotation properties are formed from the class names of the two entities to be linked. For example, the annotation property for linking the instances of the classes "Project" and "Publication" has the name "field_project_publication", and the inverse property is named "field_publication_project".

All list formats (the format name ends with "list", e.g. "node list") allow the specification of multiple values. The order of the values may be important, for example for authors of a publication. The order is represented in the ontology by "annotation of annotation", i.e., annotating the corresponding annotation (for example "field_author") using the property "ref_num" and the specification of the sequence number.

Another important function of the DOG is the generation of the directory structure for storing files (for example, datasets, images, applications, etc.) to be imported into the LHA. The DOG proceeds as follows. For each project, a directory is generated that contains a subdirectory for each of all related instances (i.e., all publications, records, methods, etc.). The subdirectories, for their part, are divided into "public" and "private". All directories are generated only if they were not already present. The DOG also links the ontology with the directory structure (under duo:File). If a file exists in one of the directories when the directory structure is generated, the DOG creates an instance of the corresponding directory class in the ontology and annotates it with the file path.

## 9.5 Simple Ontology Loader in Drupal (SOLID)

The content management system (CMS) Drupal facilitates the creation of web content (nodes) by providing simple web forms. Additionally, it allows the annotation of content with terms of self-defined vocabularies. Different node types can be defined and may be provided with fields. Fields serve as containers for the information of a concrete node. The fields support simple data types such as character strings or numbers, as well as complex types such as files and references to other nodes or vocabulary terms. However, large amounts of content to be managed can yield a complex interconnection of nodes and terms. Therefore, ontologies would be suitable for the modelling of content.

In order to enable users to import ontologies into Drupal, we have developed the Drupal module Simple Ontology Loader in Drupal (SOLID) [11]. SOLID supports both, ontologies generated by the Drupal Ontology Generator (DOG) as well as any other standard ontology (e.g., downloaded from BioPortal). Ontologies merely have to be integrated into the Drupal Upper Ontology (DUO). The module is PHP-based and interacts directly with the Drupal API. Hence, the created content does not lead to collisions or inconsistencies in Drupal's Database Management System.

SOLID is based on Drupal's module architecture and must be installed in Drupal (Version 8) to be functional. The module is accessible from the administration section of

Drupal and provides a small form for data upload and configuration to simplify the import process. Attention should be paid to the fact that nodes can only be imported if a respective node type was created prior to the import. For each property in an ontology, there must exist a corresponding field in Drupal. Automated creation of fields is not supported by SOLID because the required configuration parameters for each field are too extensive to add them into an ontology. It is much more appropriate to use the user interface provided by Drupal to create the fields. Regarding the LHA instance, we had to create the node types "project", "publication", "clinical dataset" and so forth with their respective fields (as described in Sect. 9.2). Additionally, Drupal provides the functionality to manage files (e.g., data sets or applications). Prior to the importing the ontology however, these files must be placed on the server according to the properties of the respective duo:File instance.

Hereafter, the structure and functionality of SOLID are described briefly. The module contains two types of components: parsers and importers (Fig. 9.5). Parsers are responsible for the processing of uploaded input files. OWL and JSON are supported, but this section will focus on the import of the OWL ontologies. Importers (node- respective vocabulary-importer) interact with the Drupal API, to check for existing entities and to create new ones.

In the LHA pipeline, SOLID receives the Portal Ontology of LHA (POL) from the DOG as an OWL file. The file is processed by the OWL parser (based on EasyRDF [12]). The parser extracts each subclass of the duo:Vocabulary_Concept and transmits them to the vocabulary importer. The importer creates a vocabulary in Drupal for each direct subclass of duo:Vocabulary_Concept and it adds all descending classes as terms into the vocabulary. In this step subclass/superclass relations are preserved and saved as hierarchies. Depending on the configuration, the OWL parser searches for instances of respective subclasses of duo:Node_Item in the ontology. Besides standard properties of nodes such as title, node type and alias, the parser also collects data, object and annotation properties and delivers all found properties to the node importer, where all nodes are inserted

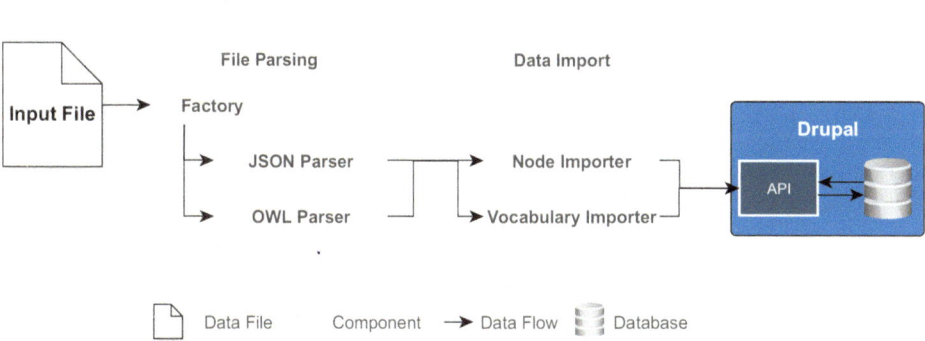

**Fig. 9.5** Architecture of the Simple Ontology Loader in Drupal

into the CMS (Fig. 9.6). In case a property references another entity in the ontology, the respective field cannot be inserted into the database immediately, because referenced nodes are potentially not yet processed and created. Therefore, referencing properties are processed after all nodes are created.

Drupal uses a Universally Unique Identifier (UUID) for a bijective identification of content which is stored in the database. To guarantee a connection between nodes after import and their source entities in the ontology, we use the entities Internationalized Resource Identifiers (IRI) as UUIDs in Drupal. By this means, the module can determine if a class or individual which is extracted by the parser already exists in the database. In

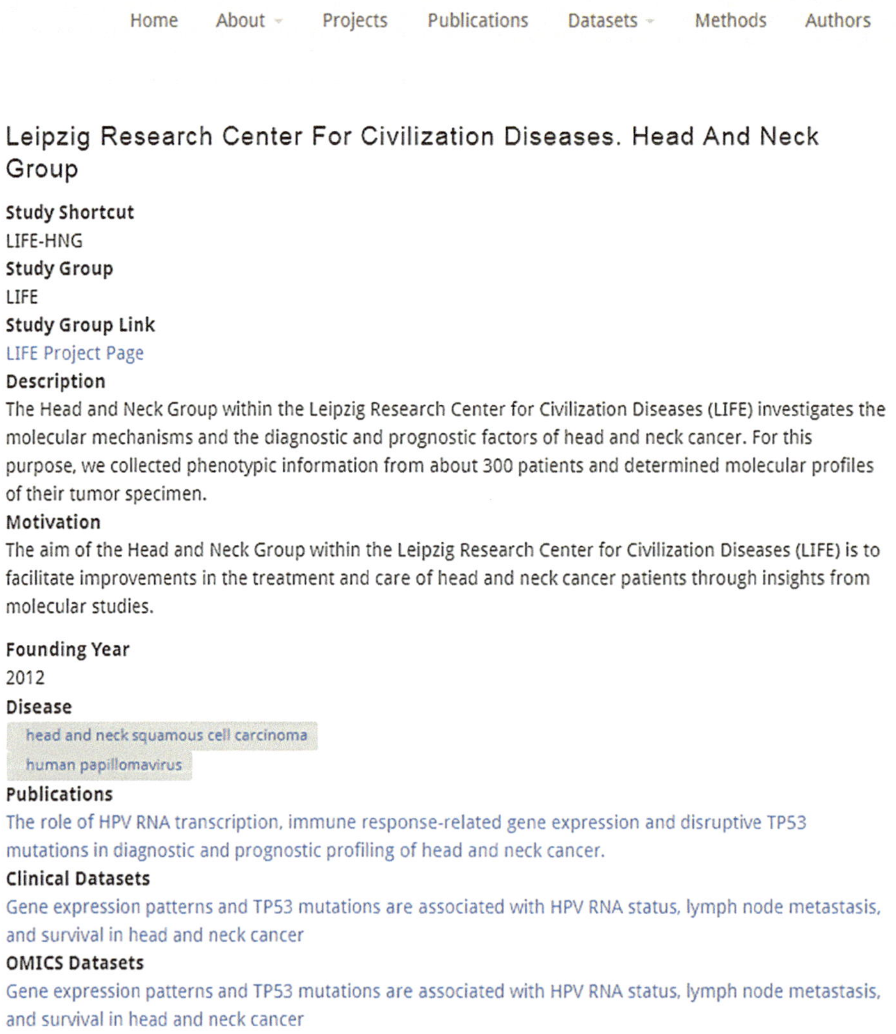

Home    About    Projects    Publications    Datasets    Methods    Authors

## Leipzig Research Center For Civilization Diseases. Head And Neck Group

**Study Shortcut**
LIFE-HNG
**Study Group**
LIFE
**Study Group Link**
LIFE Project Page
**Description**
The Head and Neck Group within the Leipzig Research Center for Civilization Diseases (LIFE) investigates the molecular mechanisms and the diagnostic and prognostic factors of head and neck cancer. For this purpose, we collected phenotypic information from about 300 patients and determined molecular profiles of their tumor specimen.
**Motivation**
The aim of the Head and Neck Group within the Leipzig Research Center for Civilization Diseases (LIFE) is to facilitate improvements in the treatment and care of head and neck cancer patients through insights from molecular studies.

**Founding Year**
2012
**Disease**
head and neck squamous cell carcinoma
human papillomavirus
**Publications**
The role of HPV RNA transcription, immune response-related gene expression and disruptive TP53 mutations in diagnostic and prognostic profiling of head and neck cancer.
**Clinical Datasets**
Gene expression patterns and TP53 mutations are associated with HPV RNA status, lymph node metastasis, and survival in head and neck cancer
**OMICS Datasets**
Gene expression patterns and TP53 mutations are associated with HPV RNA status, lymph node metastasis, and survival in head and neck cancer

**Fig. 9.6** Generated page example in the Leipzig Health Atlas web portal

case the ontological entity was imported earlier, the former node is expanded by a new revision, which contains the new fields.

The described utilization of SOLID requires the use of the web form for upload and configuration to simplify the import process. But it is also possible to direct the module via command line, to e.g., create a periodic import. New spreadsheet files may be placed in a directory of the server's file system so that DOG can create an OWL file which SOLID can subsequently import.

## 9.6 Recommendations

Our generic approach offers a solution to two kinds of problems:

1. *Development of web portals*

Our method is usable for the development of web portals with the focus on a suitable representation of structured knowledge. The following criteria should be satisfied for the development of a web portal based on our approach:

- Different types of entities having certain properties should be represented.
- There are different relationships between the particular entities.
- The entities to be represented should be annotated with concepts of terminologies/ ontologies to simplify the search.
- The content to be represented is dynamic.

In this case, the entities, their properties and relations are modelled using a spreadsheet, transformed into OWL by the DOG, and loaded in Drupal by the SOLID.

Our approach is not suitable for representing static or one-dimensional content (such as blogs), or for creating portals that require complex program logic or interaction with the user (such as forms).

2. *Representation of existing ontologies*

The number of ontologies which are available for widespread domains is growing steadily. BioPortal alone embraces over 500 published ontologies with nearly eight million classes. In contrast, the vast informative content of these ontologies is only directly intelligible by experts. To overcome this deficiency, it could be possible to represent ontologies as web portals, which do not require knowledge of ontologies and their semantics, but still, carry as much information as possible to the end-user [11]. Using our approach, ontological entities are presented to the user as discrete pages with all appropriate properties and links (to internal or external pages and files).

## 9.7    Conclusion

In this chapter, we presented an approach for specifying and automatically loading the contents of web portals into the CMS Drupal. Our approach has successfully been applied in building the LHA portal ([13], the layout of the portal is still under development), which makes available metadata, data, publications and methods from various research projects at the University of Leipzig. Ontologies have proven to be a suitable tool for modelling complex contents of web portals. Our pipeline facilitates the specification of the content by domain experts and replaces the manual input of the data in Drupal by an automated import.

Our method is generic. On the one hand, it enables the development of web portals with the focus on suitable representation of structured knowledge. On the other hand, it makes it possible to represent existing ontologies in such a way that their content is intelligible for users without background knowledge about underlying ontological entities and structures (e.g., the distinction between concepts, individuals, relations, etc.). The representation of ontological entities as traditional web pages and links facilitates access to the semantic information and improves the usability of the ontologies by domain experts.

To import an existing domain ontology into Drupal using SOLID, only a few relatively simple modifications are required. To avoid errors during the import process, some restrictions and requirements for the ontology design needed to be defined. The ontology has to be embedded in DUO, i.e., their classes and properties have to be derived from those of the DUO. Only classes and properties that are defined in DUO and are specialized or instantiated in the domain ontology are processed by SOLID. The classes whose instances are to be represented as web pages (nodes) have to be defined as subclasses of duo:Node_Item, while the root nodes of the external terminologies have to be placed under duo:Vocabulary_Concept. All annotation properties have to be subproperties of the duo:field, and their names have to match the names of the fields created in Drupal.

Our approach is a promising solution for the development of complex web portals. Additionally, it can be applied to make existing ontologies available. Future applications should be established and evaluated in further projects.

## References

1. Horgan RP, Kenny LC (2011) "Omic" technologies: genomics, transcriptomics, proteomics and metabolomics. Obstet Gynaecol 13(3):189–195
2. ISO 14721:2012. Space data and information transfer systems – Open archival information system (OAIS) – Reference model. https://www.iso.org/standard/57284.html
3. MEDLINE/PubMed XML data elements. https://www.nlm.nih.gov/bsd/licensee/data_elements_doc.html
4. Gene Expression Omnibus (GEO). https://www.ncbi.nlm.nih.gov/geo/
5. Hanauer DA, Rhodes DR, Sinha-Kumar C, Chinnaiyan AM (2007) Bioinformatics approaches in the study of cancer. Curr Mol Med 7(1):133–141(9)

6. Cerami E, Gao J, Dogrusoz U, Gross BE, Sumer SO, Aksoy BA, Jacobsen A, Byrne CJ, Heuer ML, Larsson E, Antipin Y, Reva B, Goldberg AP, Sander C, Schultz N (2012) The cBio cancer genomics portal: an open platform for exploring multidimensional cancer genomics data. Am Assoc Cancer Res. https://doi.org/10.1158/2159-8290.CD-12-0095

7. Grossman RL, Heath AP, Ferretti V, Varmus HE, Lowy DR, Kibbe WA, Staudt LM (2016) Toward a shared vision for cancer genomic data. N Engl J Med 375:1109–1112. https://doi.org/10.1056/NEJMp1607591

8. Hoehndorf R, Ngomo A-CN, Herre H (2009) Developing consistent and modular software models with ontologies. In: Fujita H, Marik V (eds) New trends in software methodologies, tools and techniques: proceedings of the Eighth SoMeT_09. Volume 199. IOS Press, pp 399–412. [Frontiers in Artificial Intelligence and Applications]

9. Herre H, Heller B, Burek P, Hoehndorf R, Loebe F, Michalek H (2006) General formal ontology (GFO): a foundational ontology integrating objects and processes. Part I: basic principles (Version 1.0). Onto-Med report. Research Group Ontologies in Medicine (Onto-Med), University of Leipzig

10. Herre H (2010) General formal ontology (GFO): a foundational ontology for conceptual modelling. In: Poli R, Healy M, Kameas A (eds) Theory and applications of ontology: computer applications. Springer, Dordrecht, pp 297–345

11. Beger C, Uciteli A, Herre H (2017) Light-weighted automatic import of standardized ontologies into the content management system Drupal. Stud Health Technol Inform 243:170–174

12. Humfrey N. RDF library for PHP. http://www.easyrdf.org/

13. Leipzig Health Atlas (LHA). https://www.health-atlas.de/

# Personalised Clinical Decision Support for Cancer Care

# 10

Bernhard G. Humm and Paul Walsh

**Key Statements**
1. Medical consultants face increasing challenges in keeping up-to-date with the rapid development of new treatments and medications, particularly for rare and complex cases.
2. Personalised medicine offers substantial benefits for patients and clinicians, but requires capturing, integrating and interpreting huge volumes of data.
3. Information providers offer evidence-based medical information services, continuously taking into account new publications and medical developments. The use of such medical information services in today's clinical day-to-day routine is, however, still limited. Due to high workload, consultants simply find no time for researching those knowledge bases. The knowledge bases themselves can also provide contradictory information.

This work was funded by the European Commission, Horizon 2020 Marie Skłodowska-Curie Research and Innovation Staff Exchange, under grant no 644186 as part of the project SAGE-CARE (SemAntically integrating Genomics with Electronic health records for Cancer CARE). Paul Walsh is supported by Science Foundation Ireland Fellowship grant number 16/IFA/4342.

B. G. Humm (✉)
Hochschule Darmstadt, Darmstadt, Germany
e-mail: bernhard.humm@h-da.de

P. Walsh
NSilico Life Science, Dublin, Ireland
e-mail: paul.walsh@nsilico.com

4. To allow effective clinical decision support, personalized medical information must be available at the point-of-care: useful information on diagnosis or treatment, tailored to each particular patient, without placing the burden of research on the consultant.

5. Personalized clinical decision support requires electronic health records to be semantically linked with evidence-based medical knowledge services.

## 10.1  Introduction

*Personalised medicine*, also known as precision or stratified medicine, is the practice of tailoring of patient care based on their predicted response or risk of disease [18, 19]. For example, patients with the same broad diagnosis, such as breast cancer, can have varying forms of the disease, such as 'Papillary Carcinoma of the Breast' or 'Ductal Carcinoma In Situ', and will have major differences in their treatment plans. When the myriad potential comorbidities, medications, symptoms, environmental factors and demographics of patients are considered, there are even more differences in optimal care paths for patients. Furthermore, since a genetic component is present in most diseases, the use of genome sequencing will allow an even more personalised approach to treatment.

There are substantial benefits for patients and clinicians in using a personalised approach as it will provide a more accurate diagnosis and specific treatment plan, offer better care outcomes for patients, will improve efficiency for the healthcare providers and will allow drug and diagnostics designers more effective methods to target disease ([18, 19]).

For example, breast cancer patients are tested for certain genetic biomarker mutations in the BRCA1 and BRCA2 genes if they have a family history of breast cancer or ovarian cancer. Patients who have pathogenic variants of BRCA1 or BRCA2 genes are at high risk for breast and ovarian cancer and prophylactic surgery if often carried out to protect their long-term health [1]. The level of BRCA1 gene expression is also an important guide for tailoring chemotherapy [2]. Note, however, that not all genetic variants are pathogenic, so the he challenge here is the appropriate clinical classification of variants and keeping this classification up-to-date, along with smooth integration of this knowledge into precision medicine.

Another example of information used in personalised medicine is the HER2 gene which contains the genetic instructions on how to make HER2 proteins, which are receptors on breast cells. HER2 proteins control how a breast tissue cell grows, divides, and repairs itself, however, in some patients, an error in the HER2 gene causes it to replicate itself leading to uncontrolled growth in the breast tissue [3]. Such information can be combined with other clinical data such as tumour size, lymph node status, comorbidities, lifestyle and to a lesser extent age and socioeconomic status to produce a more accurate diagnosis, prognosis and treatment [4].

To quote Hippocrates, "It is more important to know what sort of person has a disease, than to know what sort of disease a person has" [5].

However major challenges exist for enabling personalised medicine in front line patient care. For example a position paper by the European Society for Medical Oncology (ESMO) states that there is *"the daunting task of integrating and interpreting the ever-increasing volume of data and the associated information communication technology (ICT) needs, and the multiple dimensions of, and changing perspectives on, value and cost-effectiveness in personalised cancer medicine"* [6].

Indeed, medical consultants already face enormous challenges in keeping up-to-date with the rapid development of new treatments and medications, particularly for rare cases. Healthcare systems have undergone significant transformation over the last decade, so that clinicians are now burdened with rapidly growing medical knowledge bases, taxing regulatory requirements, increased clerical burden with paper records, the roll out of fragmented electronic health systems and intense scrutiny of quality performance indicators.

Moreover the introduction of *electronic health records (EHR)* appear to have increased clerical burden for clinicians and can distract some clinicians from meaningful interactions with patients. A recent study involving observation of 57 clinicians over many weeks indicated that clinicians spend about half of their time completing administrative tasks and interfacing with the EHR [7].

A recent study discovered that usability issues, lack of interoperability, and poor quality of documentation were sources of frustration for physicians who engage with EHR systems and it has been suggested that dissatisfaction with EHR is a major factor in physician burnout. Dissatisfaction is also trending upwards with a growing percentage of clinicians becoming frustrated with current implementations of EHRs.

Moreover, the complexity of patient data is growing and physicians are now tasked with interpreting vast volumes of patient data in the form of EHRs that aggregate documents related to a given patient. Due to the heterogeneity and volume, such data is considered as one of the most complex in the information processing industry. Biomedical big data, in the form of EHRs and digital image archiving is rapidly growing, with an estimated annual growth rate of 48% and it is estimated that health data will 2,000 exabytes or 2 zetta bytes by 2020 [8]. Consequently the management and analysis of EHR data increasingly needs big data software management tools.

Information providers offer evidence-based medical information services, continuously taking into account new publications and medical developments. Prominent examples are up-to-date (www.uptodate.com) and DynaMed Plus (www.dynamed.com).

The use of such medical information services in today's clinical day-to-day routine is, however, still limited. Due to high workload, consultants simply find no time for researching those knowledge bases. To allow effective clinical decision support, personalized medical information must be available at the point-of-care: useful information on diagnosis or treatment, tailored to each particular patient, without any research effort by the consultant.

In this chapter we describe a Clinical Decision Support System (CDSS) for cancer care [20]. The CDSS semantically links EHRs with external evidence-based medical information services, which enables the consultant to use those services without any research effort.

## 10.2   User Interaction Concept

We illustrate the user interaction concept of the CDSS by means of an example in melanoma treatment. Figure 10.1 shows anonymised summary data of a fictitious melanoma patient from an EHR.

In this example, the patient suffers from melanoma *in situ* at stage IB, with a Breslow thickness of 0.8 mm. Based on the EHR data, without interaction by the consultant, relevant information can be retrieved and displayed.

In order to provide an intuitive way for physicians and other health professionals to access this information, the CDSS leverages several information services that each try to satisfy different information needs. Those information services are organized in web page panels that the users can customize and fit to their needs by deciding which service panels should be displayed and which should be hidden. Additionally the order and size of the panels can be adapted to the user's individual needs while the resulting layout is persisted over different sessions for the specific user.

In the following sections, we briefly describe the individual information services.

### 10.2.1   Drug Information Service

One important piece of information in patient care is material on drugs and their interactions "at the point of drug prescribing" [9]. A drug information service provides information normally available in medication package leaflet inserts and secondary

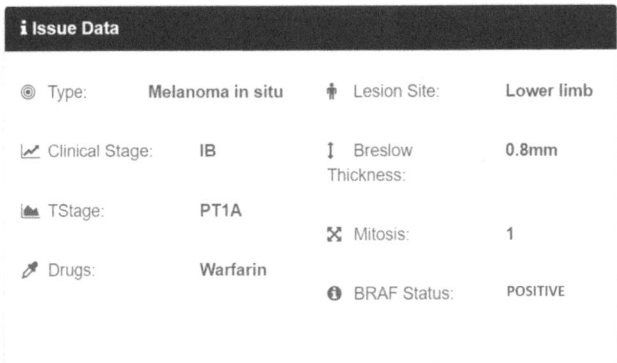

**Fig. 10.1**  Fictitious patient-related issue data

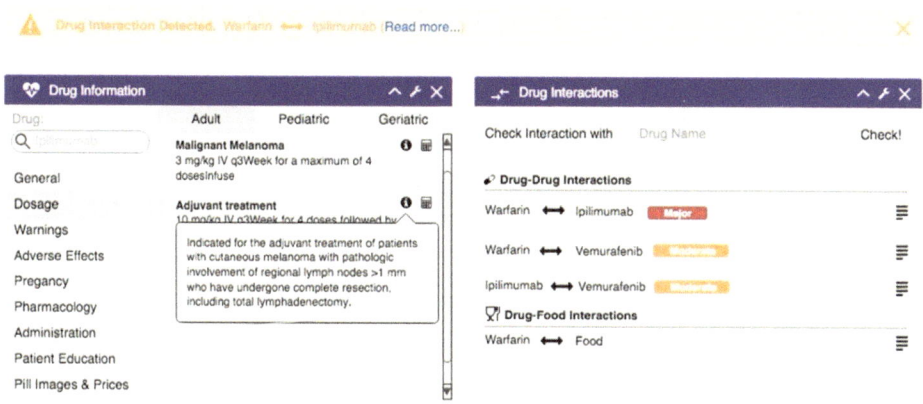

**Fig. 10.2**   Drug information service [20]

decision support services in a more accessible and structured way (Fig. 10.2, left). The provided information includes dosage data for different age groups and pre-filled calculators to suggest dosage based on relatively static information like the age and weight of the patient. For dosages dependant on more dynamic data such as renal function, it is important to acquire recent readings if the latest data is not available in the EHR. Other information available from drug information services consists of warnings, adverse effects, pharmacology, administration guidelines, material for patient education and pill images and prices. Selecting a drug to display can be done in an autosuggest-supported field that ranks already prescribed medication higher, but allows also searching for medication not yet prescribed.

As physicians indicated they wanted to see automatically generated alerts for severe drug interactions and adverse effects [9], an alert is displayed prominently (Fig. 10.2, top). For more information on how to manage the interaction or alternative drugs, an appropriate link is provided. Drug interactions as well as drug-food interactions are displayed in an another panel where the users have the possibility to check interaction with other, not yet prescribed drugs (Fig. 10.2, right).

### 10.2.2  Literature Service

The literature service displays relevant primary medical literature and review results that are related to the patient at hand (Fig. 10.3).

Automatically generated filters allow to quickly navigate the literature search results. The filters are displayed on the left whereas the medical literature is shown on the right. For each medical publication, its title, journal and publication date is displayed. In the context of evidence-based medicine (EBM), publications with a high degree of evidence are to be preferred in patient care [10]. As such, publications that are reviews or clinical trials are shown with a marker indicating their publication type. This also aligns with a

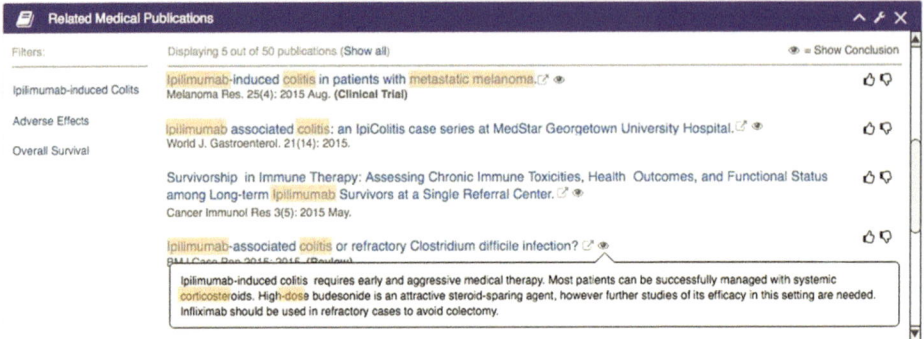

**Fig. 10.3** Literature service [20]

study from 2013 that logged and analysed data queries in a hospital and found that almost a third of the articles accessed were reviews [11, 12]. For quick orientation and relevance assessment, terms that appear in the patient's EHR are highlighted in the literature service. To help the literature relevance assessment process, a teaser text is displayed when hovering the mouse pointer over the eye icon after each publication title. In order to give the users a way to provide feedback on the relevance of a publication and improve the literature search, icons with a thumbs-up and a thumbs-down are provided.

### 10.2.3 EBM Recommendations

In this service, evidence-based medical (EBM) recommendations are displayed which are relevant for the patient currently being treated. In the example from Fig. 10.1, the patient suffers from melanoma in situ at stage IB, with a Breslow thickness of 0.8 mm. Based on the EHR data, without interaction by the consultant, the relevant page of the NCCN EBM guideline [21] for melanoma treatment is retrieved. In this example the NCCN guidelines are used, but the source of these guidelines may be configured to comply with the regulations used in a particular medical practice.

The guideline is structured as a decision tree with the relevant path (Stage IB, Breslow thickness 0.67–1.0 mm) displayed. Appropriate diagnosis and treatment procedures are recommended. Terms matching the EHR, e.g., interferon, are highlighted. If interested, the consultant may read footnotes and follow hyperlinks for more details.

### 10.3  Information Providers in Medicine

There is a large number of information providers in medicine. Some are public institutions such as the US National Institute of Health which provide information for free. Others are commercial, such as Wolters Kluwer. The volume and quality of information provided varies. Some information providers offer application programming interfaces (API) for accessing data from an EHR applications, but others provide only web access.

The following Tables 10.1, 10.2, and 10.3 give an overview of some prominent information providers.

## 10.4   Ontology-Based Electronic Health Record

Personalized clinical decision support requires EHRs to be semantically linked with evidence-based medical knowledge services. A large amount of EHR data is stored in free-text, providing the maximum flexibility for consultants to express case-specific issues. However, using free-text has a downside for mining EHR data, as medical terminology is used in diverse ways by various medical professionals and across different regions. For example, synonyms are in widespread use in the medical community, along with abbreviations, and even misspellings. While this usually poses no problem for the human expert, it is difficult for software modules to deal with those kinds of ambiguities in a reliable way.

To cope with these issues, text mining approaches have been proposed to disambiguate texts in EHRs [14]. While such an analytic approach is unavoidable when dealing with existing legacy EHR data, we use a constructive approach for new EHR applications: a semantic auto-suggest service (see Fig. 10.4).

While typing input into a free text field, suggestions of medical terms of various categories (anatomy, symptom, disease, etc.) are being presented. For example: "ipilimumab (medication)" is suggested to the user while typing "ip". While moving the mouse over an entry, an explanatory text is shown. These terms are based on merged ontologies from multiple sources, which are stored in CSV format and can be configured to suit the clinical use case.

Semantic auto-suggest not only improves usability by reducing typing effort for the consultant, but, just as importantly, it normalises the usage of medical terminology: instead of using synonyms, abbreviations or even misspelling terms, the same preferred term is always used for a concrete medical concept.

We identified six distinct semantic categories for a melanoma application: medication, activity, symptom, disease, gene, and anatomy. Some free-text fields require words from one category only, e.g., "medication used by the patient". Others can be filled with terms from multiple categories, e.g., "other relevant health issues". See Fig. 10.5.

Grounding terms used in the EHR in an ontology is the basis for semantically matching an EHR with information sources like EBM guidelines, see Fig. 10.6. The ontologies used are discussed in more detail below.

Medical terms entered into an EHR are linked to the ontology. Such terms as well as numerical data are extracted from the EHR of a particular patient. The extracted information can be used for semantically retrieving information sources like EBM guidelines which match the conditions of this patient. The relevant information is then displayed to the consultant who is making decisions on patient treatment.

**Table 10.1** Drug information sources. (Adapted from [20])

| Name | Description | API | Access | Drug information | Drug interaction | Adverse events | Drug announcements/ recalls |
|---|---|---|---|---|---|---|---|
| DailyMed | Website by U.S. National Library of Medicine (NLM), provides high quality and up-to-date drug labels. Updated daily by FDA. Documents use structured XML format | yes | public & free | ✓ | ✓ | ✓ | |
| MedlinePlus Connect | Service by NLM. provides unstructured natural language drug information/labelling and health topic overviews | yes | public & free | ✓ | | ✓ | |
| Medscape | Many clinical information resources available over website or mobile app. Articles updated yearly | no | free, registration required | ✓ | ✓ | ✓ | |
| RxNav | Provides access to different drug resources like RxNorm, NDF-RT and DrugBank Drug normalisation over different codes and systems by using RxNorm, drug interactions from DrugBank | yes | public & free | | ✓ | | |
| Wolters Klnwer Clinical Drug Information | Commercial drug information APIs including interaction, adverse effects, indications. and mapping to RxNorm | yes | commercial | ✓ | ✓ | ✓ | |

**Table 10.2** Literature information sources. (Adapted from [20])

| Name | Description | API | Access | Size |
| --- | --- | --- | --- | --- |
| Google Scholar | Search engine for scientific publications of all fields. Automatically crawls many journals | no | commercial | estimated at 160 million articles |
| Ovid | Science search platform that includes many databases, including MEDLINE | ? | subscription | ? |
| PubMed | Search engine mainly accessing MEDLINE database and focused on health topics. Query expansion by use of MeSH ontology | yes | public & free | >24.6 million records, about 500,000 new records each year |
| ScienceDirect | Website with access to large database of scientific publicatione from many fields | yes | free (abstracts), subscription (full-text) | 12 million records from 3,500 journals and 34,000 eBooks |
| Scopus | Database with abstracts and citations from many academic journals and many scientific fields, not focused on health topics | yes | paid subscription | ~55 million records |
| Springer API | Acccss to all Springer published journals, also includes BioMedCentral open-aceess publications | yes | partly free, partly subscription | ~2,000 journals and >6,500 books per year, access to >10 million online documents |

## 10.5   Ontologies in Medicine

In the medical domain, numerous controlled vocabularies, thesauri and ontologies exist. They contain medical terms and, potentially, additional information such as explanations, synonyms, hyperonyms (broader terms), and domain-specific term relationships. Following Liu and Özsu's Encyclopedia of Database Systems [16], we use the term "ontology" within this article to refer to all kinds of classified terminology in the medical domain.

Whereas some medical ontologies are commercial (e.g., Unified Medical Language System® Metathesaurus®, SNOMED-CT, etc.), there are many open source ontologies available (for an overview see, e.g., www.ontobee.org).

A challenge that needs to be addressed is how to select an ontology or a set of ontologies as the base vocabulary for the EHR application and map these ontologies to the knowledge requirements of the EHR. In analyzing the melanoma use case, we observed

**Table 10.3** EBM information sources. (Adapted from [20])

| Name | Description | API | Access | Volume |
|---|---|---|---|---|
| BMJ Best Practice | Evidence-based information to offer step-by-step guidance on diagnosis, prognosis, treatment and prevention | yes | subscription | undisclosed |
| DynaMedPlus | Evidence-based clinical overviews and recommendations. Content, updated daily. Also offers calculators, decision trees and unit and dose converters | yes | subscription | >3,200 topics and >500 journals |
| EBMeDS | Platform-Independent web service CDSS with EBM module | yes | commercial | undisclosed |
| Medscape/ eMedicine | Largest clinical knowledge base available freely. Articles updated yearly. Also available as mobile application | no | free, registration required | ˜6,800 articles |
| NCCN | Guidelines for treatment of cancer by site offering decision trees. Compiled by panels of experienced medicians | no | free, registration required | ˜60 documents |
| Physician Data Query | Cancer database from the U. S. *National Cancer institute*. Contains peer-reviewed information on cancer treatment in the form of summaries for patients and professionals | no | public | Only cancer domain |
| UpToDate | Popular evidence-based POC tool for a wide range of disciplines but targeted on internal medicine. Extensive peer-review process to ensure accurate and precise recommendations | yes | subscription, some articles free | ˜8,500 topics |

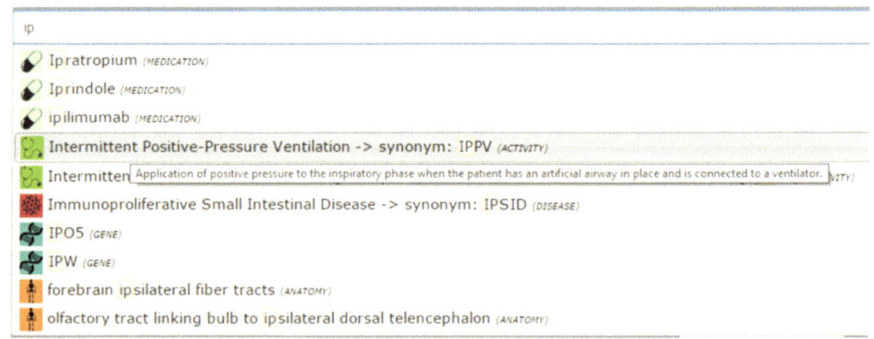

**Fig. 10.4** Semantic auto-suggest [15]

that no single ontology contains all relevant terms, i.e., necessary for the semantic auto-suggest feature. Therefore, we had to integrate several ontologies in order to get a sufficiently comprehensive ontology. For an overview of the ontologies selected, see Table 10.4 for a selection of medical ontologies and their use for different semantic categories.

| Prominent EHR field | Category and Icon | |
|---|---|---|
| medication used by the patient | Medication | |
| treatments the patient received | Activity | |
| diagnoses and findings by the physicians | Symptom | |
| patients's diseases | Disease | |
| findings of the gene analysis | Gene | |
| body parts where a melanoma occurs | Anatomy | |
| other relevant health issues | all above | |

**Fig. 10.5**   Semantic categories [15]

## 10.6   Software Architecture

### 10.6.1 Overview

Figure 10.7 gives an overview of the CDSS software architecture.

The architecture is separated into an online subsystem and an offline subsystem. The offline subsystem is a batch process for integrating various source ontologies into an application-specific ontology. It is implemented as a semantic extract, transform, load (ETL) process. The online subsystem is organized as a three-layer-architecture consisting of client, business logic and data store. See also ([17]; [22]).

Components with semantic logic are the semantic ETL, ontology services and decision support services. In the following sections, we describe some aspects of the semantic components. For more details see ([13, 15, 20]).

### 10.6.2 Semantic ETL

For integrating various ontologies into an application-specific ontology, e.g., used for semantic auto-suggest, data needs to be extracted from the source ontologies, transformed, and loaded into a data store (ETL). The following issues need to be addressed:

1. *Transformation of technical data formats*: Ontologies have different technical formats, e.g., XML, XLS, CSV, RDF. A transformation from the specific to the common format is required.
2. *Semantic field mapping*: Even if the technical formats are identical, e.g., XML, the individual field names and structure of the ontologies may differ. For example, broader terms in MeSH are encoded as tree id whereas in other ontologies, the ids of the broader terms are listed.
3. *Semantic cleansing/filtering*: Some terms are "polluted" (have unwanted parts) or are not meaningful for the semantic application. For example, the general term

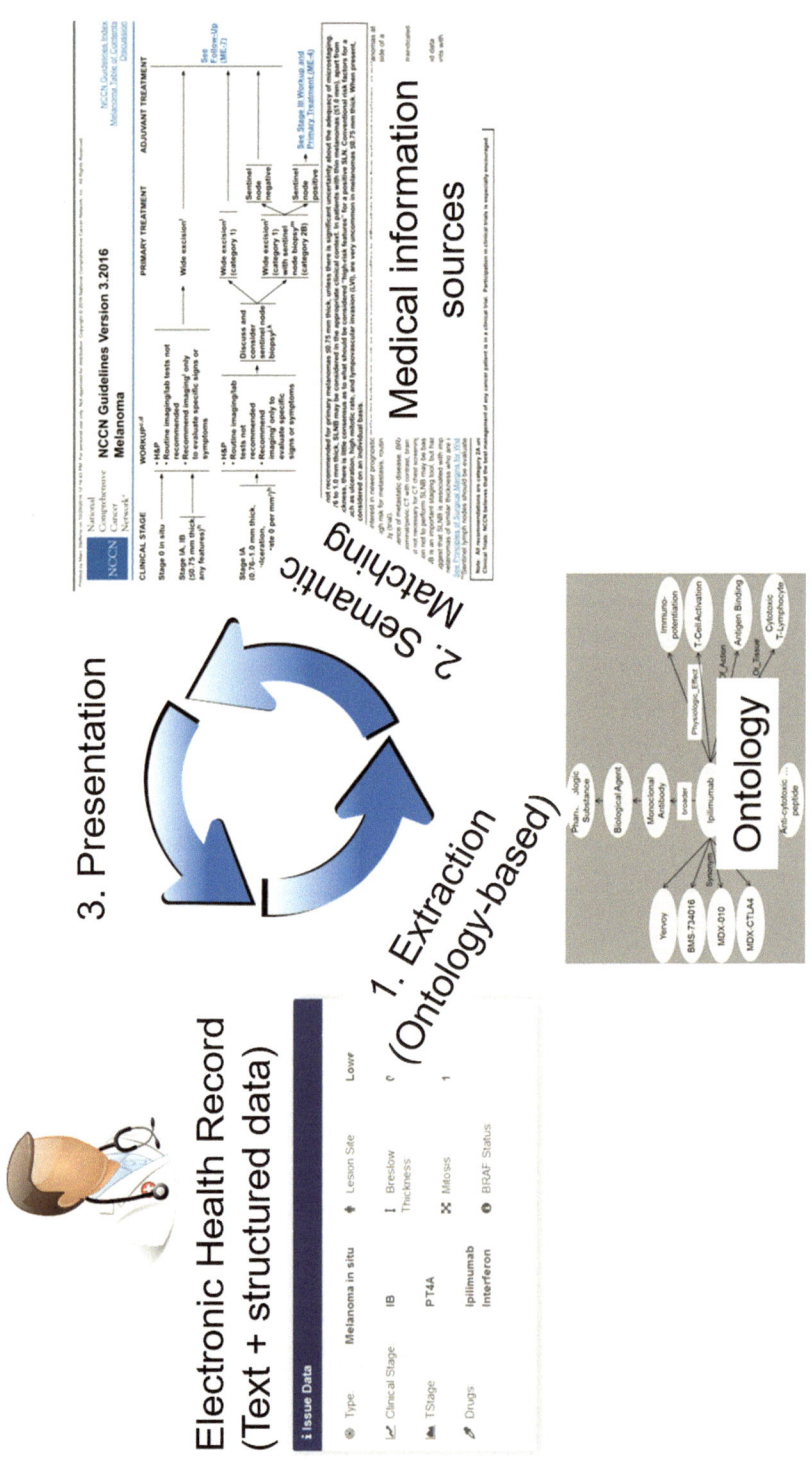

**Fig. 10.6** Semantically matching EHRs with medical information sources

**Table 10.4**   Medical ontologies for various semantic categories

| Name | Anatomy | Symptom | Gene | Disease | Activity | Medication | License |
|---|---|---|---|---|---|---|---|
| The Drug Ontology (DRON) | | | | | | x | open |
| National Drug File Reference Terminology (NDF-RT) | | | | | | x | open |
| Human disease ontology (DOID) | | | | x | | | open |
| Anatomical Entity ontology (AEO) | x | | | | | | open |
| Foundational Model of Anatomy (FMA) | x | | | | | | open |
| Uber anatomy ontology (UBERON) | x | | | | | | open |
| Gen Ontology (GO) | | | x | | | | open |
| Ontology of Genes and Genomes (OGG) | | | x | | | | open |
| VIVO-ISF | | | | | x | | open |
| Symptom Ontology (SYMP) | | x | | | | | open |
| Medical Subject Headings (MeSH) | x | x | x | x | x | x | Registration necessary |
| NCI Thesaurus (National Cancer Institute) | x | x | x | x | x | x | open |

"Non-physical anatomical entity" from the Foundational Model of Anatomy does not denote a concrete body part. Those terms need to be filtered out.

4. *Duplicate handling*: Duplicate terms occur because some terms are covered in various ontologies (e.g., "Warfarin" is covered in The Drug Ontology and in MeSH), and even in various versions within the same ontology. Duplicates need to be removed.

5. *Target data format and storage*: The target ontology data format as well as the data store technology used should be targeted towards the intended application. E.g., for semantic auto-suggest a simple data format consisting of term, semantic category, definition, hyponyms, and synonyms suffices. A search index, such as Apache Solr, provides optimal performance and allows semantic search of terms, their category, hierarchy of hyponyms as well as synonyms.

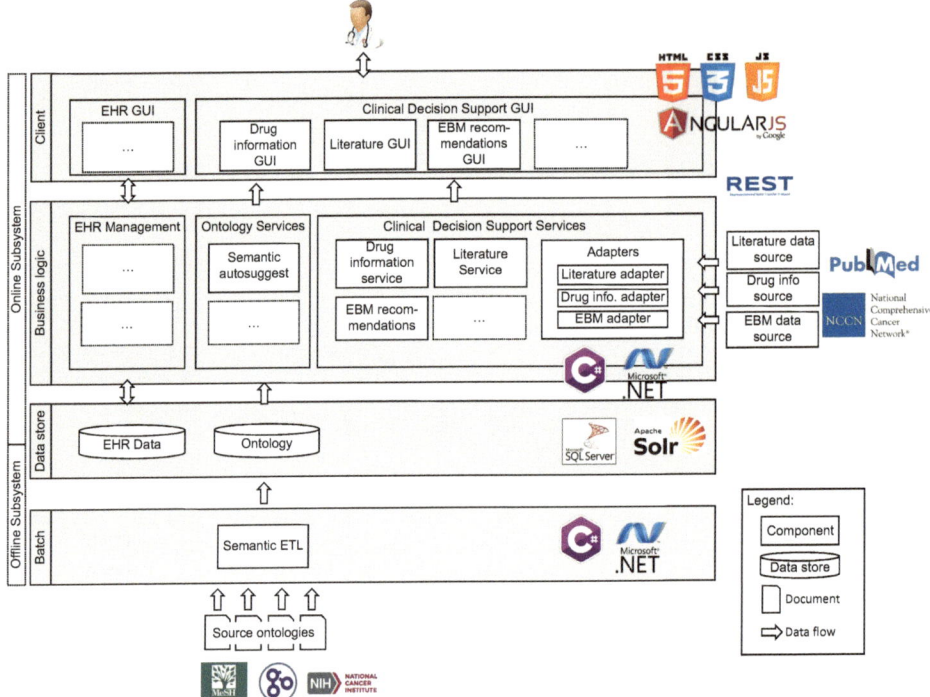

**Fig. 10.7** CDSS software architecture

### 10.6.3 Literature Service

For selecting literature relevant for the patient's treatment, relevant data needs to be extracted from the EHR and used for querying literature data sources such as PubMed. The semantic matching logic is application-specific; specific to the medical specialty, specific to the EHR management application, and specific to the literature data source.

See Fig. 10.8 for a query generation example from a melanoma EHR for PubMed.

From the ca. 100 attributes that are used in the EHR application, not all are helpful for personalized literature suggestions. Fields with no relevance, such as the patient's name, are omitted in query generation, whereas relevant fields like the issue, medications or comorbidities are included. The query to be generated must conform to the query language of the data source selected, here, PubMed. The query itself is generated by a rule-based template engine. For example, one rule to search for publications that address the safety or efficacy aspects of the medications prescribed, combines all medications with an "OR" and adds "(safety OR efficacy)" to the subquery. Another rule combines the comorbidities field with the medication to search for drug-related adverse effects and their treatment. To ensure data quality and only search for recent literature, restrictions are added to the query, such as the "hasabstract[text]" to only show publications that contain an abstract.

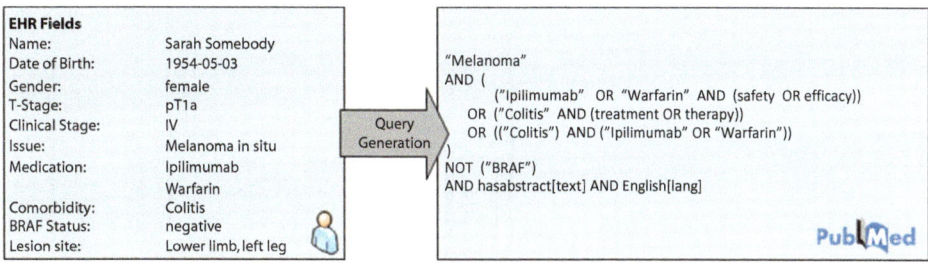

**Fig. 10.8**   Sample query generation from an EHR for PubMed [20]

### 10.6.4 EBM Recommendation Service

Identifying sections of EBM guidelines which are relevant for a particular patient under treatment requires more than full text search. Consider the patient example above (patient data in Fig. 10.1). The EBM data source chosen are NCCN guidelines, here for melanoma, which are provided as a PDF document. The task is to identify the one section in a 150 page document which exactly matches the patient's condition. In the example above, the patient's Breslow thickness is 0.8. Searching for the string "0.8" in the text of the NCCN guidelines will not match the relevant page (ME-3), since on this page, the condition is formulated as "0.67–1.0 mm thick". Therefore, some explicit decision logic is required for matching extracted EHR data to sections of the EBM guidelines. See Fig. 10.9 for an example rule.

Here, the following rule is shown: "If clinical stage is IB and Breslow thickness is between 0.76 and 1.0 mm, then section ME-3 on page 8 is relevant". This rule is edited using a business rule composer, here MS BizTalk [23].

Applying the rules in a business rule engine with the extracted EHR data as input will match the relevant section of the EBM guidelines which can then be displayed to the consultant in the clinical decision support system. Using a business rule composer may have advantages over coding the decision logic in a conventional programming language. It allows adding or modifying business rules by trained medical administrators whenever new or modified EBM guidelines are published. Where possible metadata such as author, reputation, affiliations, version and time of publication can be used to provide confidence to the users on the accuracy of the presented guidelines.

### 10.6.5 Implementation

We have successfully implemented the personalised clinical decision support system for melanoma care. The offline subsystem and the business logic has been implemented in C# using Microsoft .Net technology. We use Microsoft SQL server for storing EHRs and Apache Solr for storing and querying the ontology for the semantic auto-suggest

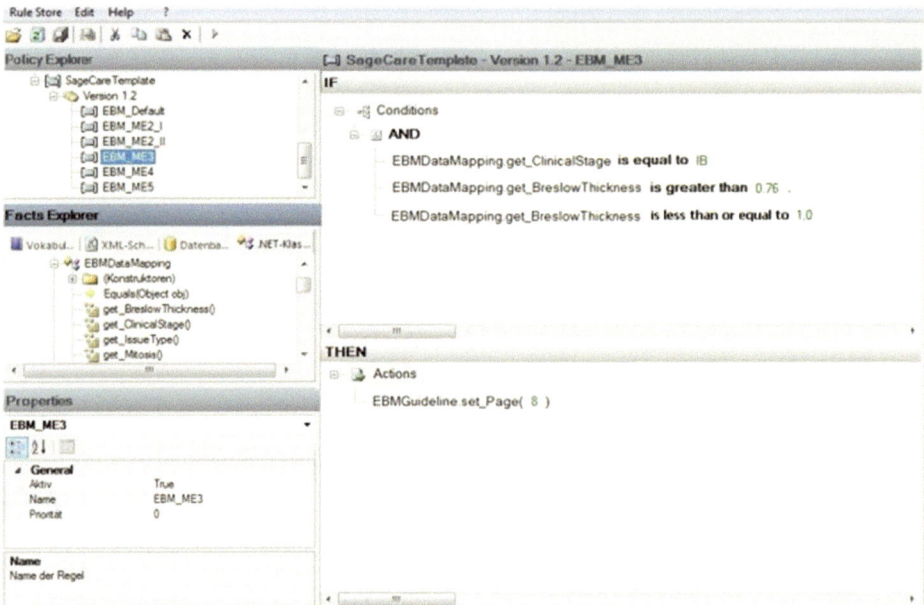

**Fig. 10.9** Example rule (MS Biztalk Server) [13]

service. The client accesses the server via a REST interface. The client is implemented in HTML5/CSS/JavaScript respectively Type Script, using Google's AngularJS framework. See Fig. 10.7.

## 10.7 Recommendations

We summarize our main learnings from implementing the personalised clinical decision support system in the following recommendations.

1. When developing a semantic application, carefully look at at regulatory compliance to check what are the constraints on selecting a particular ontology, then check for existing ontologies which suit the application use case. Pay more attention to regulatory compliance and the quality and completeness of data than to technical data formats used. In the medical domain, there are numerous ontologies available.
2. Carefully analyze the quality of ontologies with respect to the application use case. It is most common that no single existing ontology meets the quality requirements of an application use case and can be used without modification.

3. Use Semantic ETL for preprocessing existing ontologies for an application use case. It includes the process steps extraction, transformation, semantic cleansing/filtering, and loading.
4. Ontologies provide a common terminology within a semantic application and may be used for mapping to the terminology of an information provider service.
5. When including services of information providers in semantic applications, carefully check suitability for the application use case, technical constraints, and license details.
6. Semantically mapping EHRs with clinical information sources requires application-specific code, taking into account the specifics of the medical specialty, the EHR application, and the information source.

## 10.8 Conclusion

Personalised medicine promises many benefits. However, it is not yet in widespread use in clinical practise. We believe that personalised medicine must seamlessly integrate into the consultant's workflow, posing no additional workload for searching relevant medical information. In this article we have presented a personalised clinical decision support system for cancer care, which provides the consultant treating a patient with relevant medical information based on the patient's EHR.

We have successfully implemented the clinical decision support system. After successful tests it is planned to be integrated into a commercial EHR application.

## References

1. Petrucelli N, Daly MB, Pal T (1998) BRCA1- and BRCA2-associated hereditary breast and ovarian cancer [Online]. Available at https://www.ncbi.nlm.nih.gov/pubmed/20301425. Accessed 1 Dec 2017
2. Papadaki C, Sfakianaki M, Ioannidis G, Lagoudaki E, Trypaki M, Tryfonidis K, Mavroudis D, Stathopoulos E, Georgoulias V, Souglakos J (2012) ERCC1 and BRAC1 mRNA expression levels in the primary tumor could predict the effectiveness of the second-line cisplatin-based chemotherapy in pretreated patients with metastatic non-small cell lung cancer. J Thorac Oncol 7(4):663–671
3. breastcancer.org (2017) HER2 status [Online]. Available at http://www.breastcancer.org/symptoms/diagnosis/her2. Accessed 1 Dec 2017
4. Soerjomataram I, Louwman MWJ, Ribot JG, Roukema JA, Coebergh JW (2008) An overview of prognostic factors for long-term survivors of breast cancer. Breast Cancer Res Treat 107(3):309–330
5. Murugan R (2015) Movement towards personalised medicine in the ICU. Lancet Respir Med 3(1):10–12

6. Ciardiello F, Arnold D, Casali PG, Cervantes A, Douillard J-Y, Eggermont A, Eniu A, McGregor K, Peters S, Piccart M, Popescu R, Van Cutsem E, Zielinski C, Stahel R (2014) Delivering precision medicine in oncology today and in future – the promise and challenges of personalised cancer medicine: a position paper by the European Society for Medical Oncology (ESMO). Ann Oncol 25(9):1673–1678
7. Shanafelt TD, Dyrbye LN, West CP (2017) Addressing physician burnout: the way forward. JAMA 317(9):901–902
8. Privacy Analytics (2016) The rise of big data in healthcare [online]. Available at https://privacy-analytics.com/de-id-university/blog/rise-big-data-healthcare/. Accessed 17 Dec 2017
9. Rahmner PB, Eiermann B, Korkmaz S, Gustafsson LL, Gruvén M, Maxwell S, Eichle HG, Vég A (2012) Physicians' reported needs of drug information at point of care in Sweden. Br J Clin Pharmacol 73(1):115–125
10. Hung BT, Long NP, Hung LP, Luan NT, Anh NH, Nghi TD et al (2015) Research trends in evidence-based medicine: a joinpoint regression analysis of more than 50 years of publication data. PLoS ONE 10(4):e0121054. https://doi.org/10.1371/journal.pone.0121054
11. Maggio LA, Cate OT, Moorhead LL, Van Stiphout F, Kramer BM, Ter Braak E, Posley K, Irby D, O'Brien BC (2014) Characterizing physicians' information needs at the point of care. Perspect Med Educ 33(5):332–342
12. Maggio LA, Steinberg RM, Moorhead L, O'Brien B, Willinsky J (2013) Access of primary and secondary literature by health personnel in an academic health center: implications for open access. J Med Libr Assoc 101(3):205–212
13. Humm BG, Lamba F, Landmann T, Steffens M, Walsh P (2017) Evidence-based medical recommendations for personalized medicine. In: Proceedings of the collaborative european research conference (CERC 2017), Karlsruhe
14. Jensen PB, Jensen LJ, Brunak S (2012) Mining electronic health records: towards better research applications and clinical care. Nat Rev Genet 13(6):395–405
15. Beez U, BG Humm, Walsh P (2015) Semantic autosuggest for electronic health records. In: Arabnia HR, Deligiannidis L, Tran Q-N (eds) Proceedings of the 2015 international conference on computational science and computational intelligence. IEEE Conference Publishing Services, Las Vegas, 7–9 Dec 2015. ISBN 978-1-4673-9795-7/15, https://doi.org/10.1109/CSCI.2015.85
16. Liu L, Özsu MT (eds) (2009) Encyclopedia of database systems. Springer, New York
17. Humm BG, Walsh P (2015) Flexible yet efficient management of electronic health records. In: Arabnia HR, Deligiannidis L, Tran Q-N (eds) Proceedings of the 2015 international conference on computational science and computational intelligence. IEEE Conference Publishing Services, Las Vegas, 7–9 Dec 2015. ISBN 978-1-4673-9795-7/15, https://doi.org/10.1109/CSCI.2015.84
18. Academy of Medical Sciences (2015a) Stratified, personalised or P4 medicine: a new direction for placing the patient at the centre of healthcare and health education (Technical report). Academy of Medical Sciences. May 2015
19. Academy of Medical Sciences (2015b) Stratified, personalised or P4 medicine: a new direction for placing the patient at the centre of healthcare and health education [Online]. University of Southampton Council; Science Europe; Medical Research Council. Available at https://acmed-sci.ac.uk/viewFile/564091e072d41.pdf. Accessed 1 Dec 2017
20. Idelhauser J, Beez U, Humm BG, Walsh P (2016) A clinical decision support system for personalized medicine. In: Bleimann U, Humm B, Loew R, Stengel I, Walsh P (eds) Proceedings of the 2016 European collaborative research conference (CERC 2016), Cork, pp 132–145. ISSN 2220-4164

21. National Comprehensive Cancer Network (2017) Online. https://www.nccn.org/. Last Accessed 12 Jan 2017
22. Coym M, Humm BG, Spitzer P, Walsh P (2017) A dynamic product line for an electronic health record. In: Bleimann U, Humm B, Loew R, Regier S, Stengel I, Walsh P (eds) Proceedings of the collaborative European research conference (CERC 2017), Karlsruhe, pp 134–141, 22–23 Sept 2017. ISSN 2220-4164
23. Microsoft BizTalk Server (2017) Online. https://www.microsoft.com/en-us/cloud-platform/biztalk. Last Accessed 12 Jan 2017

# Applications of Temporal Conceptual Semantic Systems

**11**

Karl Erich Wolff

**Key Statements**

1. The challenging problem of describing and understanding multidimensional temporal data in practice can often be solved as shown in this article. We describe an application of a mathematical temporal theory in chemical industry where the behavior of a distillation column has to be understood with respect to many variables.
2. The mathematical theory employed is Formal Concept Analysis (FCA) and its temporal extension Temporal Concept Analysis (TCA). It offers the possibility of representing the semantic meaning of data with respect to the chosen aspect of the expert technician.
3. Those aspects can be visualized in planar diagrams of multidimensional conceptual structures where each of the selected variables is represented in a suitable granularity.
4. More general than other temporal theories, TCA offers a broadly applicable notion of a *state* of a *temporal object* at a *time* in a certain *view*. A view represents the chosen aspect of the expert.
5. The most valuable tool in TCA is the representation of trajectories in multidimensional diagrams which helps the expert technician to understand the dynamics of the distillation column with respect to seven variables simultaneously.

K. E. Wolff (✉)
Hochschule Darmstadt, Darmstadt, Germany

© Springer-Verlag GmbH Germany, part of Springer Nature 2018                     145
T. Hoppe et al. (eds.), *Semantic Applications*,
https://doi.org/10.1007/978-3-662-55433-3_11

## 11.1    Introduction

### 11.1.1 Semantic Scaling

The main purpose of this chapter is to present the broad applicability of a general semantic strategy which I call *semantic scaling*. The main idea in semantic scaling is to explain used terms, e.g. abbreviations or values, in a broader setting by interpreting the meaning of these terms with respect to some purpose. Since many descriptions in industrial or scientific practice are often reduced to a data table, we focus here on semantic scaling of values in data tables. Allowing not only numbers but arbitrary terms as values of data tables we consider these terms as concepts, hence as basic notions for describing relational statements in the form of a sequence of concepts. That opens the way to describe temporal relational phenomena in any applications.

In many applications, in practice, the values in a given data table have a meaning for the specialist which is often hidden for others. For example the velocity value of "100 miles/hour" might be associated with the attribute of "dangerous". There may even be relational meanings of such values, for example, the meaning of "100 miles/hour" might be "If somebody is driving a car in a town travelling at least 100 miles/hour, it might lead to dangerous situations". In the following, we simplify evaluation of data by attaching just attributes – and not more complicated relational descriptions – to the values of each variable. This is a simple and effective method to generate a suitable granularity for the intended data evaluation. It is demonstrated by an example from chemical industry where temporal data of a distillation column has to be evaluated.

### 11.1.2 Semantic Scaling of Temporal Data of a Distillation Column

In a chemical firm, the process in a distillation column had to be investigated with respect to 13 variables such as *input*, *pressure* and *reflux*. For each of the 13 variables at each of 20 days, at most one value had been measured. For six variables missing values occurred. In this paper we focus on the seven variables without missing values and evaluate them. For the other six variables, the same procedure can be applied. A typical part of the data is shown in Table 11.1. The main *problem* was to understand the dynamics of this distillation column with respect to many variables. Obviously, several temporal questions about the description of states of the distillation column and its frequency in a given state, as well as dependencies between variables should be investigated.

In cooperation with specialists of this distillation column, the author has applied semantic scaling for each of the 13 variables. For example, the variable *input* varies between 600 and 675, while the values of *pressure* vary between 100 and 130. To construct valuable insights for the specialists, their understanding of the employed variables had to be represented, for example, their understanding of important regions in the domain of each

**Table 11.1** Temporal data of a distillation column

| day | input | pressure | reflux | energy1 | ... | variable 13 |
|-----|-------|----------|--------|---------|-----|-------------|
| 1 | 616 | 119 | 129 | 616 | ... | ... |
| 2 | 603 | 125 | 174 | 680 | ... | ... |
| 3 | 613 | 118 | 133 | 629 | ... | ... |
| ... | ... | ... | ... | ... | ... | ... |
| 15 | 639 | 116 | 174 | 588 | ... | ... |
| ... | ... | ... | ... | ... | ... | ... |
| 20 | 664 | 120 | 127 | 556 | ... | ... |

**Table 11.2** A scale with two attributes for the *input* values

| input | ≤615 | ≤645 |
|-------|------|------|
| 600 | X | X |
| 601 | X | X |
| 602 | X | X |
| ... | ... | ... |
| 639 | | X |
| ... | ... | ... |
| 675 | | |

variable. In a linear (or one-dimensional) domain, regions can be easily described by bounds, for instance to separate normal from dangerous regions. However, in two or even higher dimensions, the specialists often have only vague or no idea which regions might be important for their purpose. We shall show in this paper how states of the distillation column can be visualized in multidimensional spaces, to aid the specialists in interpreting these multidimensional aspects. To explain that, we start by introducing some attributes for the input values as shown in Table 11.2. This table serves in this paper as an example of a *formal context* and particularly as a *conceptual scale*, the main tool of semantic scaling in Formal Concept Analysis.

The first column contains all integers from 600 to 675 inclusive, covering the full range of measured input values for the corresponding variable. The two attributes "≤615" and "≤645" had been chosen after discussion with the specialists of the distillation column to provide a coarser understanding of the distribution of the measured input values together with other similarly scaled variables. The numerical information that $600 \leq 615$ is indicated by a cross "X" in the row of 600 and the column of "≤ 615"; the other crosses have the corresponding meaning. It is obvious that Table 11.2 just divides the set of all integers from 600 to 675, [600, 675], into two subsets, namely the set [600, 615] and the set [600, 645]. Clearly, $[600, 615] \subseteq [600, 645] \subseteq [600, 675]$. We shall see that this chain of three sets is the set of *extents* of the *concept lattice* of the *formal context* given by Table 11.2. To explain our way to represent the temporal data in Table 11.1 we give a short introduction to Formal Concept Analysis.

### 11.1.3 Formal Concept Analysis

Formal Concept Analysis (FCA) is a mathematical theory originating from the three basic theories in mathematics, namely logic, geometry and algebra. Their ordinal structures have been generalized by G. Birkhoff [3] in his book on "Lattice Theory". It was used for classification purposes by M. Barbut and B. Monjardet [1]. R. Wille [6] realized the connection between lattice theory and the philosophical understanding of the concept of "concept". Since philosophers often start from some basic notions of "objects" and "attributes" and use the binary relation that "an object has an attribute", R. Wille introduced the mathematical definition of a *formal context* (G,M,I) where G and M are sets and I is a binary relation between G and M, $I \subseteq G \times M$. When $(g,m) \in I$ we say "g has the attribute m", written gIm. Set G is called the set of formal objects (German: Gegenstände), set M is called the set of formal attributes (German: Merkmale), and set I is called the set of incidences. Clearly, small formal contexts can be represented as cross tables, as for example in Table 11.2.

R. Wille introduced the notions *formal concept* and *concept lattice* for a given formal context. There are computer programs for generating the concept lattice of a finite formal context and interactive programs for the generation of a graphical representation of a concept lattice in form of a *line diagram*.

In FCA, semantic scaling is done for the values of each field (column) of a given data table by constructing a formal context for that field, called a conceptual scale. In a conceptual scale, all values of the scaled field are taken as formal objects. They are described by suitable attributes with respect to the purpose of the investigation. A conceptual scale can be constructed to be information preserving or it can focus on some special partial knowledge. From a scaled data table, i.e., all fields are scaled, we construct a formal context, the derived context of the scaled data table. It combines the objects measured in the data table with the attributes describing the measured values. This technique of semantic scaling a data table and constructing its derived context is called *conceptual scaling*. For the mathematical definition the reader is referred to [5].

For understanding temporal data, Formal Concept Analysis has been extended by the author to Temporal Concept Analysis (TCA) by introducing notions of temporal objects, temporal concepts, views, and a general definition of a state of a temporal object at a time in a preselected view [7–11].

## 11.2 Conceptual Scaling of Temporal Data of a Distillation Column

In the following, we apply conceptual scaling to the data in Table 11.1. Roughly speaking, for each variable (e.g., *input*) we use a formal context, called a conceptual scale of this variable (e.g., Table 11.2). The conceptual scale represents both a semantic meaning and simultaneously a granularity (as determined by an expert) for the values of the variable. To combine the meaning of the values with the temporal variable day, we replace each value in Table 11.1 by its corresponding row in the conceptual scale of the variable and obtain the derived context. For example, to obtain the derived context, $K_i$, for a subset of Table 11.1

with only the variable *input*, we replace the values in the table with the corresponding row in Table 11.2, to obtain the derived context as given in Table 11.3. As names for the attributes in the derived context we take (input, ≤615) and (input, ≤645).

When we apply conceptual scaling also to the variable *energy1* using the scale attributes "≤570" and "≤630" we obtain the following derived context $K_{ie}$ in Table 11.4.

Using Table 11.4 as a typical result of conceptual scaling, we now explain the mathematical core of conceptual scaling. For that purpose we first mention that in Table 11.1 at day = 1 the input is 616, for short: input(1) = 616. Since 616 does not have the scale attribute "≤615" there is no cross in Table 11.4 in the cell for day = 1 and the attribute "(input, ≤615)". Since 616 has the scale attribute "≤645" there is a cross in Table 11.4 in the cell for day = 1 and the attribute "(input, ≤645)". To lead the reader to the general definition of the derived context, we introduce the standard notation for the given example. Let g be an arbitrary formal object of the given data table (in our example g = 1), and let m be an arbitrary attribute in the given data table (in our example m = input) and let n be an arbitrary scale attribute of the scale of m (in our example n = (≤615): then the formal objects of the derived context are, by definition, the formal objects of the given data table (in our example the set {1, 2, …, 20}), the formal attributes are by definition the pairs (m,n) where m is an attribute in the given data table and n is a scale attribute of the scale of m with its incidence relation $I_m$. Then the incidence relation of the derived context is denoted by J and defined by

$$g\,J\big(m,n\big):\Leftrightarrow m\big(g\big)I_m\,n,$$

In words: in the derived context, a formal object g has the attribute (m,n) if and only if the value m(g) has the attribute n in the scale of m.

**Table 11.3** The derived context $K_i$ for the *input* variable scaled with Table 11.2

| day | (input, ≤615) | (input, ≤645) |
|-----|-----|-----|
| 1 |     | X |
| 2 | X | X |
| 3 | X | X |
| … | … | … |
| 15 |     | X |
| … | … | … |
| 20 |     |     |

**Table 11.4** The derived context $K_{ie}$ for *input* and *energy1*

| day | (input, ≤615) | (input, ≤645) | (energy1, ≤570) | (energy1, ≤630) |
|-----|-----|-----|-----|-----|
| 1 |     | X |     | X |
| 2 | X | X |     |     |
| 3 | X | X |     | X |
| … | … | … | … | … |
| 15 |     | X |     | X |
| … | … | … | … | … |
| 20 |     |     | X | X |

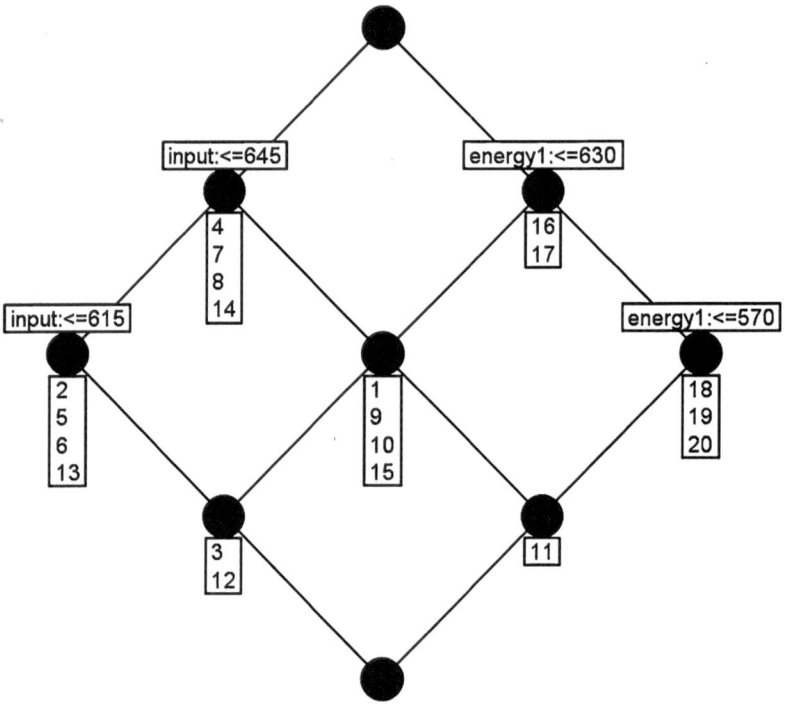

**Fig. 11.1** Concept lattice of the derived context $K_{ie}$ in Table 11.4

In our example that reads: 1 J (input, $\leq$615) $\Longleftrightarrow$ input (1) $I_{input} \leq$615 .

In words: In the derived context the formal object g = 1 is incident with the attribute (input, $\leq$615) if and only if the input(1) has the scale attribute $\leq$615 in the input scale. In this example input(1) = 616 does not have the scale attribute $\leq$615. Hence for day = 1 there is no cross in the derived context at the attribute (input, $\leq$615).

We shall see later that the formal context $K_{ie}$ can be reconstructed from the concept lattice in Fig. 11.1.

In Fig. 11.1 all formal objects {1, 2, ..., 20} of the derived context $K_{ie}$ occur in the labels under the circles and all attributes of the $K_{ie}$ occur in the labels above the circles. Whether a formal object has an attribute or not can also be seen from the concept lattice in Fig. 11.1. This will be explained in the following section.

## 11.3   The Concept Lattice of a Formal Context

In FCA a concept lattice is understood as a hierarchy of formal concepts constructed from a formal context (G,M,I). Each formal concept is a pair (A,B) where A is a subset of the set G, and B is a subset of M satisfying some condition given later. Then A is called the *extent* and B the *intent* of (A,B).

Figure 11.1 shows the *concept lattice* of the formal context $K_{ie}$ given in Table 11.4. Before explaining *formal concepts* and *concept lattices* we give a coarse description of Fig. 11.1.

## 11.3.1 Examples of Concepts, Object Concepts and Attribute Concepts

Each circle in Fig. 11.1 represents a formal concept of $K_{ie}$. For example the circle labeled by the days 1,9,10,15 denotes the formal concept (A,B) where the extent $A = \{1,9,10,15\} \cup \{3,12\} \cup \{11\}$ $= \{1,3,9,10,11,12,15\}$ and the intent B = {(input, <=645), (energy1, <=630)}.

For each circle one may find the represented formal concept roughly speaking as follows: Find the extent by searching downwards for formal objects, find the intent by searching upwards for attributes.

For a given formal concept (A,B) each formal object in A has each attribute in B, and (A,B) is maximal with respect to this condition. This is often expressed by: Each formal concept forms a rectangle full of crosses in a cross table where rows and columns are suitably permuted. Extent or intent may be the empty set.

For example, in Fig. 11.4 the circle at the top denotes the formal concept $(\{1,...,20\}, \emptyset)$, since there is no attribute in $K_{ie}$ satisfying all formal objects. The circle at the bottom denotes the concept

$(\emptyset, \{(\text{input}, <= 615), (\text{input}, <= 645), (\text{energy1}, <= 570), (\text{energy1}, <= 630)\}),$

since there is no formal object in $K_{ie}$ having all attributes.

One of the most important properties of concept lattices is that each concept lattice contains all the information of its formal context. To explain that we introduce the object concepts which will play a prominent role in Temporal Concept Analysis.

For a given formal context (G,M,I) and any $g \in G$, the set of all attributes of g, namely $g^\uparrow := \{m \in M \mid g \, I \, m\}$, is the intent of the object concept of g. Its extent is the set of all objects having all attributes of $g^\uparrow$, i.e. $g^{\uparrow\downarrow} := \{h \in G \mid h \, I \, m$ for all $m \in g^\uparrow\}$. The object concepts of g is defined by

$$\gamma(g) := \left(g^{\uparrow\downarrow}, g^\uparrow\right).$$

The object concepts of $K_{ie}$ are represented in Fig. 11.4 by the circles which have at least one of the labels from {1, ... 20} just below their circle.

Similarly, for any $m \in M$ the set $m^\downarrow := \{g \in G \mid g \, I \, m\}$ is the extent of the attribute concept of m. Its intent is the set of all attributes satisfied by all objects of $m^\downarrow$, i.e. $m^{\downarrow\uparrow} := \{n \in M \mid g \, I \, n$ for all $g \in m^\downarrow\}$. The attribute concept of m is defined by

$$\mu(m) := \left(m^\downarrow, m^{\downarrow\uparrow}\right).$$

The attribute concepts of $K_{ie}$ are represented in Fig. 11.4 by the circles which have at least one of the attribute labels just above their circle.

## 11.3.2 Formal Concepts, Concept Lattices and Implications

Slightly more general than the definitions of the object concepts and the attribute concepts is the general definition of a *formal concept*: Let (G,M,I) be a formal context. For any subset $X \subseteq G$ we construct the set $X^\uparrow := \{m \in M \mid g \, I \, m$ for all $g \in X\}$ of the attributes common to the objects in X. For any subset $Y \subseteq M$ the set $Y^\downarrow := \{g \in G \mid g \, I \, m$ for all $m \in Y\}$ is the set of objects which have all attributes in Y. Using this notation we can cite the famous definition of a formal concept given by Wille [6] (see also Ganter and Wille [5]):

A *formal concept* of (G,M,I) is a pair (A,B) with $A \subseteq G$, $B \subseteq M$, $A^\uparrow = B$ and $B^\downarrow = A$. A is called the *extent* and B the *intent* of (A,B).

The set of all formal concepts of (G,M,I) is denoted by **B**(G,M,I). It can be ordered by the following subconcept relation: If $(A_1, B_1)$ and $(A_2, B_2)$ are formal concepts of (G,M,I), then $(A_1, B_1)$ is called a *subconcept* of $(A_2, B_2)$ provided that $A_1 \subseteq A_2$ (which is equivalent to $B_2 \subseteq B_1$). In this case we write $(A_1, B_1) \leq (A_2, B_2)$. (We use the same sign "$\leq$" here as for the natural order relation for numbers since that should not lead to difficulties.)

The ordered structure ( **B**(G,M,I), $\leq$ ) is called the *concept lattice* of (G,M,I).

For example, in the concept lattice (**B**($K_{ie}$), $\leq$) shown in Fig. 11.1, the object concept $\gamma(3)$ of day 3 is a proper subconcept of $\gamma(1)$, shortened $\gamma(3) < \gamma(1)$, and $\gamma(1) < \mu$(input, $\leq 645$), hence $\gamma(3) < \mu$(input, $\leq 645$).

Using this example, the meaning of the lines in Fig. 11.1 can be explained easily. For example, the line from the circle of $\gamma(3)$ to the circle of $\gamma(1)$ expresses that $\gamma(3)$ is a *lower neighbour* of $\gamma(1)$ which means that $\gamma(3) < \gamma(1)$ and there is no formal concept c fulfilling $\gamma(3) < c < \gamma(1)$. Clearly, $\gamma(3)$ is not a lower neighbour of $\mu$(input, $\leq 645$).

Any formal context (G,M,I) can be reconstructed from its concept lattice since for any $g \in G$ and $m \in M$ the following *reading rule* holds:

$$g \, I \, m \iff \gamma(g) \leq \mu(m).$$

Hence we can reconstruct $K_{ie}$ from its concept lattice in Fig. 11.1, for example $\gamma(1) \leq \mu$(energy1, $\leq 630$), hence 1 J (energy1, $\leq 630$), i.e. at day 1 energy1 $\leq 630$.

In the following we use *implications* between attributes. If in a formal context (G,M,I) all formal objects satisfying an attribute m also satisfy an attribute n, we say that *m implies n*, for short m $\Rightarrow$ n. That can be expressed as $m^\downarrow \subseteq n^\downarrow$, or equivalently as $\mu(m) \leq \mu(n)$. For example, in Fig. 11.1 (input, $\leq 615$) $\Rightarrow$ (input, $\leq 645$). We shall also use implications between subsets A and B of M and say A $\Rightarrow$ B if $A^\downarrow \subseteq B^\downarrow$. For example, in Fig. 11.2 {(input, $\leq 645$), (energy1, $\leq 570$)} $\Rightarrow$ {(reflux, $\leq 140$)}. That means, that at each day where input $\leq 645$ and energy1 $\leq 570$ – there is only one such day, namely 11 – the reflux $\leq 140$. For further information about implications the reader is referred to [5], p. 79.

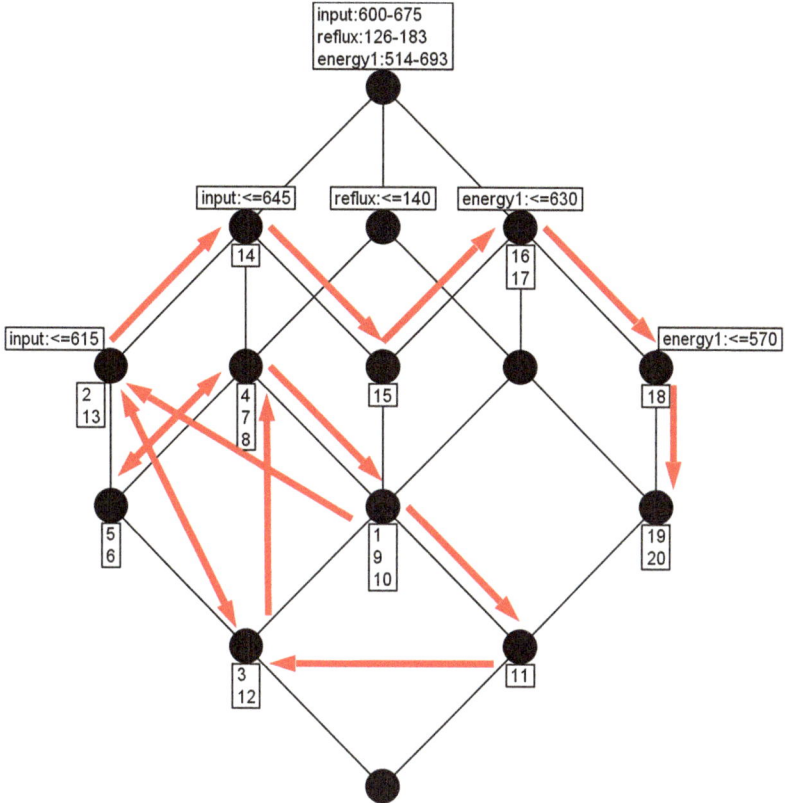

**Fig. 11.2**  A transition diagram for input, reflux and energy1

---

## 11.4   Temporal Evaluation of the Distillation Data

In Table 11.1, the attribute *day* is a key of this data table. That characterises the most simple form of temporal data. In this case, the notion of a state is easily understandable. For the general definition of a state the reader is referred to [8–10].

### 11.4.1  States of the Distillation Column

Looking at the concept lattice in Fig. 11.1 and interpreting the formal objects as days, it is obvious that the object concept of each day represents, in a natural way, what is usually called a *state*. For example, at the first day the distillation column is in the state described in Fig. 11.1 by the object concept $\gamma(1)$ with its intent {(input, $\leq$645), (energy1, $\leq$630)}. Therefore, for such temporal systems – which are described in [8–10] – a *state* is defined as an object concept of the derived context. The distillation column is – with respect to Table 11.1 and for an arbitrary scaling – at each day t in exactly one state, namely $\gamma(t)$. Hence, in that formal sense, the distillation column behaves like a particle in physics that is at each time t at exactly one place $x(t)$.

The chosen concept lattice plays the role of a state space. Only some of the formal concepts in Fig. 11.1 occur as states. For example the top concept and the bottom concept do not belong to the set of states. For any state $\gamma(t)$ the set of formal objects g where $\gamma(g) = \gamma(t)$ is called the *contingent* of this object concept. The elements of the contingent of a state are shown just below the circle of this state. The number of elements in the contingent of a state is the *frequency* of this state. For example, the frequency of $\gamma(1)$ is 4, since the contingent of $\gamma(1)$ is $\{1,9,10,15\}$.

## 11.4.2 Transitions and Trajectories

To introduce *transitions* and *trajectories* we use the natural successor relation on the integers, in our example the natural successor relation on the set of days $\{1, ..., 20\}$. For any element $(t, t+1)$ of the successor relation its *transition* is defined as the pair

$$\Big((t,t+1),\big(\gamma(t),\gamma(t+1)\big)\Big)$$

consisting of the *basic transition (t,t+1)* and its 'image' under $\gamma$. This is generalized in TCA for more general temporal systems.

In *transition diagrams*, the transition $((t,t+1), (\gamma(t), \gamma(t+1)))$ is represented as an arrow leading from the circle of $\gamma(t)$ to the circle of $\gamma(t+1)$ if $\gamma(t) \neq \gamma(t+1)$, for $t \in \{1, ..., 19\}$ in Fig. 11.2. For example, the transition $((5,6), (\gamma(5), \gamma(6)))$ is not drawn in Fig. 11.2 since $\gamma(5) = \gamma(6)$. The sequence of all transitions is called a *trajectory*. For formal definitions, the reader is referred to [10].

The concept lattice in Fig. 11.2 is slightly more complicated than that in Fig. 11.1 since we have extended the formal context $K_{ie}$ by the attribute (reflux, $\leq 140$). The three attributes shown at the top concept, tell the range of the three variables; input, reflux and energy1. We call this new derived context $K_{ire}$. It has been constructed to visualize the behavior of the distillation column with respect to the three variables input, reflux and energy1 in a coarse granularity as it is used in practical applications. Therefore, we use the terms low, middle and high for input and energy1, and low and high for reflux. Using these terms we see from the transition diagram in Fig. 11.2 that the distillation column starts with middle input, middle energy1 and low reflux. After a short excursion at day 2 to low input, high reflux and high energy1 it moves until day 12 in the region of middle or low input and simultaneously low reflux with all levels of energy1. Then it climbs up to high reflux and high energy1 at day 13, but with low input. The following part of the trajectory is quite remarkable. In five steps the distillation column moves from low to high input, from high to low energy1 and in the last

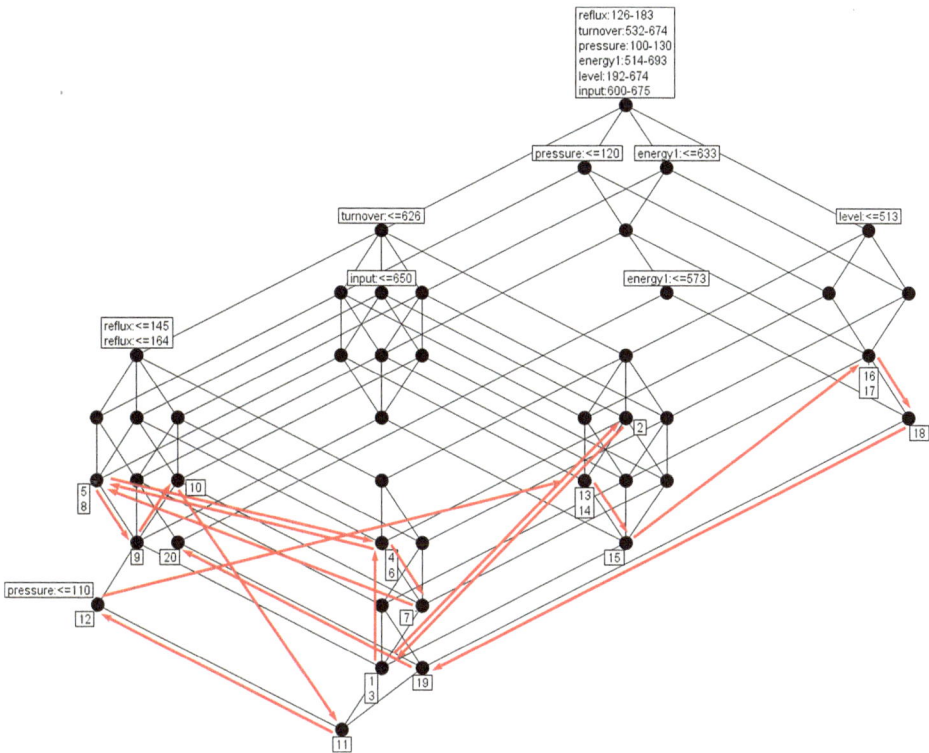

**Fig. 11.3** The trajectory of the distillation column representing six variables

step from day 18 to day 19 it changes from high to low reflux. These five last arrows follow neighbourhood edges in this concept lattice. We call these transitions *neighbourhood transitions*.

While Fig. 11.2 serves in this paper to introduce transition diagrams in an easily understandable concept lattice representing only three variables, we now show in Figs. 11.3, 11.4, and 11.5 how concept lattices can be used to understand the behavior of the distillation column in seven variables. The mathematical theory for describing temporal relational structures is based on Temporal Relational Concept Analysis where Temporal Relational Semantic Systems has been introduced by the author [9, 10].

## 11.4.3  A Conceptual Map for the Distillation Data

To construct a conceptual map for discussing and understanding the behavior of the distillation column in many variables, we proceed as follows. Since the experts of the distillation column mostly prefer to discuss the domains of the variables in three steps, namely low, middle and high we now scale all variables by ordinal scaling such that each domain is divided into three parts, as was done for *input* and *energy1* in Fig. 11.1. To keep the map small we use only the seven variables without missing values. This leads to a clearly structured concept lattice, part of which is shown in Fig. 11.3 representing six variables.

The trajectory of the distillation column is shown in a concept lattice of a derived context with the attributes shown in Fig. 11.3. At the top concept we show all six variables used in this context to visualize their domains. The line diagram in Fig. 11.3 is drawn such that for each circle the intent of its represented formal concept can be easily seen. For example, at day 16 the distillation column is in state $\gamma(16) = \gamma(17)$ satisfying the attributes (pressure, $\leq 120$), (energy1, $\leq 633$), (level, $\leq 513$) and clearly also all six attributes at the top concept which are satisfied by all states. In the state $\gamma(18)$ only a single attribute, namely (energy1, $\leq 573$), has to be added to the intent of $\gamma(17)$ to get the intent of $\gamma(18)$. The transition from day 17 to 18 is a neighbourhood transition, as well as that from 18 to 19 – but $\gamma(19)$ has three attributes more than $\gamma(18)$, namely (reflux, $\leq 145$), (reflux, $\leq 164$), (turnover, $\leq 626$). Note, that $\gamma(19)$ is not a subconcept of $\mu$(input, $\leq 650$), therefore by the reading rule, day 19 does not have the attribute (input, $\leq 650$).

Figure 11.3 shows much more, for example that all states satisfy at least one of the attributes (level, $\leq 513$), (reflux, $\leq 164$). One can also see that the attribute concept $\mu$(input, $\leq 650$) is a subconcept of $\mu$(turnover, $\leq 626$), hence (input, $\leq 650$) $\Rightarrow$ (turnover, $\leq 626$). We also see that (reflux, $\leq 164$) and (reflux, $\leq 145$) have the same attribute concept. That means that (reflux, $\leq 145$) $\Leftrightarrow$ (reflux, $\leq 164$). Obviously (reflux, $\leq 145$) $\Rightarrow$ (reflux, $\leq 164$) while (reflux, $\leq 164$) $\Rightarrow$ (reflux, $\leq 145$) means that in this context there is no day with a reflux value in the interval [146,164]. Finally we mention that {(pressure, $\leq 110$)} $\Rightarrow$ {(reflux, $\leq 145$), (input, $\leq 650$), (energy1, $\leq 633$)}. Some trivial implications are not mentioned here.

Figure 11.3 gives a remarkable and quite simply structured 'landscape' which contains nearly all value bounds introduced to divide the domain of each of six variables into three parts. But some of these bounds and the bounds for the variable *feed* are not yet included. They are shown in Fig. 11.4.

The main purpose of Fig. 11.4 is to introduce the meaning of the *nested* lattice structure in Fig. 11.5. One could use Figs. 11.3 and 11.4 to discuss and understand the behavior of the distillation column in seven variables, each scaled in a chain with three concepts, but one can combine both diagrams to a single *nested line diagram* shown in Fig. 11.5 (see [5], pp. 75–79).

In Fig. 11.5 the trajectory of the distillation column is shown in a *nested line diagram* which represents a derived context for seven variables. Each of these variables is scaled by a conceptual scale such that the concept lattice of this scale is a chain consisting of three concepts. For example, the three scale attributes used for the variable pressure are $\leq 110$, $\leq 120$ and $100$–$130$. In the derived context, the corresponding attribute concepts

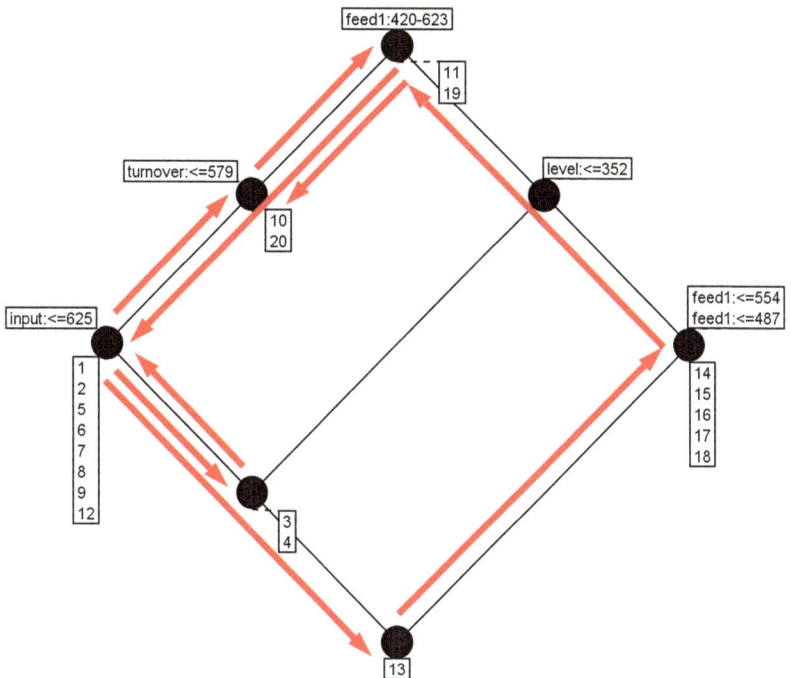

**Fig. 11.4** The attributes from the seven 3-chain scales not yet used in Fig. 11.3

occur also as a chain, namely $\mu$(pressure, $\leq$110) $\leq \mu$(pressure, $\leq$120) $\leq \mu$(pressure, 100–130). For reflux and feed, two of the corresponding attributes are equivalent as shown in Fig. 11.5.

In Fig. 11.5 each circle of Fig. 11.3 is inflated and filled with a copy of the lattice structure of Fig. 11.4. In this nested line diagram, the inflated circles together with their neighbourhood edges form the *outer* diagram while the small circles in each inflated circle form the *inner* diagram. The 15 attributes from Fig. 11.3 can be seen in the outer diagram, while the 6 attributes from Fig. 11.4 appear in the inner diagram in the top circle of the outer diagram. The object labels from 1 to 20 for describing the states appear at their corresponding object concepts. For each state $\gamma(g)$ its intent is the union of its outer intent and its inner intent. For more information about nested line diagrams the reader is referred to [5].

It is obvious that reading such complicated diagrams needs some training. But then we can use these diagrams as a 'conceptual map' which allows understanding of the behavior of the distillation column with respect to many variables in the chosen granularity.

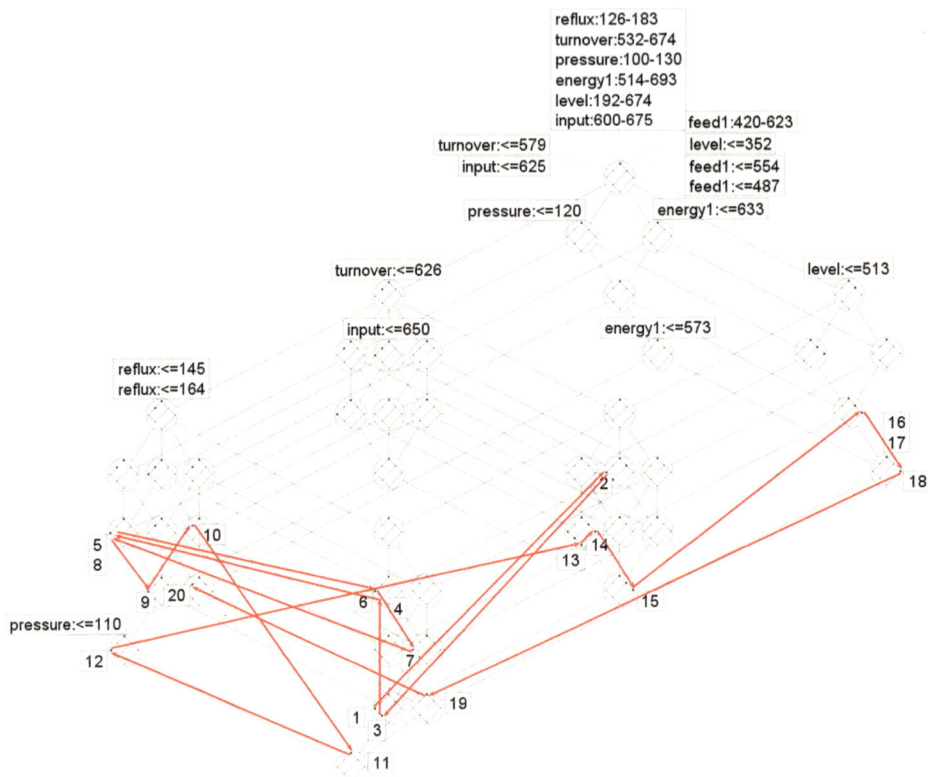

**Fig. 11.5** Trajectory in a concept lattice showing seven variables each scaled in a 3-chain

## 11.5 Discussion

### 11.5.1 Applicability

Applying *Formal Concept Analysis* has two main effects. At first, users from industry, medicine or any other field are surprised by the *general applicability* of formal concepts compared with the well-known algebraic and metric structures of numbers and vector spaces. One of the main disturbing effects is the *huge variety of unknown forms* of concept lattices. Therefore, users usually need a long time to become familiar with the lattice structures of FCA. However, when they are trained in reading and understanding concept lattices they appreciate the advantages of FCA. They understand that their data is represented with respect to their self-chosen scales such that the concept lattice of the derived context just shows the original data without any loss with respect to their chosen granularity. They also feel the close connection between their own conceptual thinking and with the mathematical structures in FCA.

Applying *Temporal Concept Analysis* has the great advantage that it offers a general and flexible notion of a *state* which is connected with the chosen granularity. It also offers a clear notion of handling multiple *temporal objects*, an aspect which was not discussed in this paper as only a single temporal object, namely the distillation column, appears in this application. TCA had been applied by the author successfully in psychoanalysis, medicine, biology and several times in industry. It generalizes the understanding of states in automata theory by connecting the notion of a state with an explicit time representation and with arbitrary granularity. Even the fine granularity of the continuum of real numbers is included in TCA. This yields a clear conceptual understanding of the notions of particles and waves in physic.

## 11.5.2 Methodology

In human communication, natural language is the main tool for knowledge representation and knowledge processing. Statements about the world use concepts. Often the meaning of the employed concepts is not clear enough for the intended purpose. To describe the meaning of concepts for a given purpose, *semantic scaling*, as briefly described in the introduction, is used. The main methodical tool in FCA is a special kind of semantic scaling which focuses on describing concepts simply by attributes instead by arbitrary relational statements. Applying FCA usually represents concepts used in practice as formal concepts in conceptual scales which describe the meaning of the concepts in a suitable granularity. These formal concepts are then connected in conceptual scaling with the given data table by the derived context which can be visualized by a line diagram of its concept lattice. The visualization can be used to understand complex data in many dimensions.

## 11.5.3 Technology

To compute concept lattices and to draw line diagrams, the following programs have been used. For scaling temporal data we use the program Cernato and export the scaled data in xml-format to the program Siena, which is a part of the ToscanaJ Suite. Siena is used to draw the line diagrams. The trajectories of temporal objects are embedded into the line diagram (or nested line diagram) using Siena's Temporal Concept Analysis tool, yielding the transition diagrams. Using the program ToscanaJ, the main program of the ToscanaJ Suite, one can search in huge conceptual systems by iterated zooming into concepts of interest. For further information the reader is referred to [2, 5]. ToscanaJ is available for download at https://sourceforge.net/projects/toscanaj/files/. To get Cernato, write to the author (karl.erich.wolff (at) t-online.de).

### 11.5.4 Experiences

The author has applied FCA in several industrial and scientific projects. During his consultancy in the mentioned chemical firm he carried out the initial steps in developing TCA. For all participants of this project, one of the most successful sessions was the common development of conceptual scales for the single variables. While in this paper we focus on ordinal scales where the concept lattice for a single variable is just a chain, we also discussed in practice several other scales such as biordinal scales for two "parallel" chains, interordinal scales for the representation of intervals in a suitable granularity, nominal scales for representing equality and inequality, Boolean scales for scaling subsets, and several combinations of scales. It was possible to generate trajectories of the distillation column in 2, 3 and 4 dimensions. These trajectories proved to be very successful for the experts of the distillation column to understand its behavior. They recognized that, prior to this, they had a partial misunderstanding of their distillation column. The diagrams generated in this chapter would give them an even better understanding of their distillation column.

Before summarizing the main learnings from applying FCA and TCA, some other projects using TCA, and therefore FCA, should be mentioned. The first project was the representation of the development of an anorectic young woman and her family over about 2 years. This example is mentioned in [8] together with the first part of the theoretical development of TCA. The next project was the investigation of the notions of 'particles' and 'waves' in physics [7]. A project in medicine and biology investigated gene expression processes and visualized the behavior of six patients in multidimensional spaces based on their genetic data [12]. In another project, the criminological investigation of conversations of pedophiles with children via chats was supported by visualizations of these data represented in temporal relational semantic systems [4].

## 11.6    Recommendations

### 11.6.1 Generality of FCA/TCA

The first recommendation is also a warning: FCA and TCA are quite general mathematical theories which have the great advantage that they can be applied successfully in many situations: finite or infinite, discrete or continuous, in sciences or in humanities, and in industry as well as in public domain. Concepts formed in natural language often can be represented successfully as formal concepts of suitably built formal contexts. Applications usually treated with statistics can easily be handled with FCA or TCA since numbers can be understood as special formal concepts. When someone wishes to apply FCA or TCA they should be willing to learn new structures.

## 11.6.2 Simplicity of Formal Contexts Versus Complexity of Concept Lattices

One of the first impressions for beginners in FCA and TCA is the simplicity of the structure of a formal context. This simplicity contrasts with the possible complexity of the concept lattice of a formal context. In addition to reading and understanding line diagrams, actually representing concept lattices in well-drawn line diagrams requires experience. Even more experience is necessary for designing suitable scales and generating views, such that the given data can be interpreted easily. Therefore, the second recommendation is: beginners in FCA/TCA should contact experts in FCA/TCA before starting their project.

## 11.6.3 Recommendations Concerning the Programs of FCA/TCA

For beginners in FCA it is meaningful to use first the program Concept Explorer (see: http://conexp.sourceforge.net/) to obtain a feeling for concept lattices. Among others it mainly contains a context editor, a drawing program for generating (non-nested) line diagrams and the interactive Implication Program. For drawing nice line-diagrams one should use Siena in the ToscanaJ Suite. Up to now, there is only a single program for TCA, namely the TCA module in Siena. It can automatically embed the trajectories of up to 10 temporal objects of a temporal system into a line diagram of some part of the derived concept lattice with respect to the chosen conceptual scaling. The scaling is done in Cernato which is a program in the NaviCon Decision Suite. My recommendation is to attend a course in FCA/TCA explaining these programs. For further advice see: http://ernstschroederzentrum.de/ and http://www.upriss.org.uk/fca/fca.html.

## 11.7   Conclusions

The application of Formal Concept Analysis (FCA) and Temporal Concept Analysis (TCA) offers deep insight into data. In this chapter the example of a distillation column and its behavior represented simultaneously in seven variables had been chosen to demonstrate some of the main ideas in FCA. It also demonstrates in a very simple temporal system two main ideas in TCA, namely the flexible granularity dependent notions of states and trajectories. With respect to applications of FCA/TCA, this chapter has demonstrated possibilities and discussed advantages and difficulties occurring in practical applications.

# References

1. Barbut M, Monjardet B (1970) Ordre et classification. Algèbre et Combinatoire. 2 tomes. Hachette, Paris
2. Becker P, Hereth Correia J (2005) The ToscanaJ suite for implementing conceptual information systems. In: Ganter B, Stumme G, Wille R (eds) Formal concept analysis. LNAI 3626. Springer, Heidelberg, pp 324–348
3. Birkhoff G (1967) Lattice theory, 3rd edn. American Mathematical Society, Providence
4. Elzinga P, Wolff KE, Poelmans J, Dedene G, Viaene S (2012) Analyzing chat conversations of pedophiles with temporal relational semantic systems. In: Domenach F, Ignatov D, Poelmans J (eds) Formal concept analysis. Contributions to the 10th international conference on formal concept analysis (ICFCA 2012). Leuven, Belgium, pp 82–101
5. Ganter B, Wille R (1999) Formal concept analysis: mathematical foundations. Springer, Heidelberg. German version: Springer, Heidelberg (1996)
6. Wille R (1982) Restructuring lattice theory: an approach based on hierarchies of concepts. In: Rival I (ed) Ordered sets. Reidel, Dordrecht/Boston, pp 445–470. Reprinted in: Ferré S, Rudolph S (eds) Formal concept analysis. ICFCA 2009. LNAI 5548. Springer, Heidelberg, pp 314–339 (2009)
7. Wolff KE (2004) 'Particles' and 'waves' as understood by temporal concept analysis. In: Wolff KE, Pfeiffer HD, Delugach HS (eds) Conceptual structures at work. LNAI 3127. Springer, Heidelberg, pp 126–141
8. Wolff KE (2005) States, transitions, and life tracks in temporal concept analysis. In: Ganter B, Stumme G, Wille R (eds) Formal concept analysis – state of the art. LNAI 3626. Springer, Heidelberg, pp 127–148
9. Wolff KE (2007) Basic notions in temporal conceptual semantic systems. In: Gély A, Kuznetsov SO, Nourine L, Schmidt SE (eds) Contributions to ICFCA 2007, Clermont-Ferrand, Laboratoire LIMOS, Université Blaise Pascal, Aubière CEDEX, pp 97–120
10. Wolff KE (2010) Temporal relational semantic systems. In: Croitoru M, Ferré S, Lukose D (eds) Conceptual structures: from information to intelligence. ICCS 2010. LNAI 6208. Springer, Heidelberg, pp 165–180
11. Wolff KE (2011) Applications of temporal conceptual semantic systems. In: Wolff KE et al (eds) Knowledge processing and data analysis. LNAI 6581. Springer, Heidelberg, pp 59–78
12. Wollbold J, Wolff KE, Huber R, Kinne R (2011) Conceptual representation of Gene expression processes. In: Wolff KE et al (eds) Knowledge processing and data analysis. LNAI 6581. Springer, Heidelberg, pp 79–100

# Context-Aware Documentation in the Smart Factory

**12**

Ulrich Beez, Lukas Kaupp, Tilman Deuschel, Bernhard G. Humm,
Fabienne Schumann, Jürgen Bock, and Jens Hülsmann

**Key Statements**

1. In factory environments, it is important to quickly identify appropriate technical documentation for machinery in error and maintenance situations.
2. Smart factory is the vision of a production environment with increasingly self-organising and self-adapting machinery. Identifying appropriate technical documentation in error and maintenance situations will become even more important in the smart factory.

---

The project ProDok 4.0 – Process-Oriented Technical Documentation for Industry 4.0 – is funded by the German Ministry of Education and Research (BMBF) within the framework of the Services 2010 action plan under funding no. 02K14A110.

U. Beez (✉) · L. Kaupp · T. Deuschel · B. G. Humm
Hochschule Darmstadt, Darmstadt, Germany
e-mail: Ulrich.Beez@h-da.de; lukas.kaupp@h-da.de; tilman.deuschel@h-da.de;
bernhard.humm@h-da.de

F. Schumann
dictaJet Ingenieurgesellschaft mbH, Wiesbaden, Germany
e-mail: fabienne.schumann@dictajet.de

J. Bock
KUKA Roboter GmbH, Augsburg, Germany
e-mail: Juergen.Bock@kuka.com

J. Hülsmann
ISRA Surface Vision GmbH, Herten, Germany
e-mail: jhuelsmann@isravision.com

3. In order to identify appropriate documentation, the semantic context of the error or maintenance situation needs to be taken into account. The semantic context needs to be extracted and inferred from low-level machine data.
4. The ProDok 4.0 application allows identifying appropriate documentation in error and maintenance situations for two use cases: robotics application development and maintenance of industrial inspection machines.

## 12.1    Introduction

In every factory environment, errors and maintenance situations may occur. They must be handled quickly and accurately. Highly experienced and skilled experts for maintenance and repair know exactly what they have to do in most situations. However, the larger and more complex the factory environment and the more specific the error situation, the more likely even experts need to consult technical documentation, not to mention less skilled workers who depend on accurate, easy-to-use technical documentation. But how is it possible quickly and easily identify appropriate technical documentation in an error or maintenance situation?

Technical documentation informs the user of a machine about how to operate it safely and in the intended way, about error situations, maintenance procedures, and proper disposal. Depending on the region, there are different regulations concerning technical documentation. Inside the European Union, technical documentation has to be delivered to the customer in accordance with the Directive 2006/42/EC of the European Parliament and of the Council on machinery [5], such as user manuals, installation and assembly instructions, and maintenance manuals. Chapter 1.7 of ANNEX I – Essential health and safety requirements relating to the design and construction of machinery – deals with information and warnings on machinery, and with information devices connected to the machinery. It is stipulated that "the information needed to control machinery must be provided in a form that is unambiguous and easily understood. It must not be excessive to the extent of overloading the operator. Visual display units or any other interactive means of communication between the operator and the machine must be easily understood and easy to use" [5].

However, first identifying and then searching for the appropriate technical documentation in a given error or maintenance situation is difficult and requires much more than a full-text search. What does the highly skilled and experienced expert do when being faced with an error? First, he will analyse the situation, taking into account the symptoms he observes. Using those observations, the expert will form hypotheses on the error cause based on this experience and will generally know what to do in order to solve the problem.

For the less skilled worker, support in performing these steps can be provided through a semantic software application. To achieve this, the application needs to extract low-level machine data in order to infer semantic context information (symptoms), link symptoms to causes and solutions semantically, and present appropriate solutions to the user in an easy-to-use way.

In this chapter, we describe such a semantic application for identifying appropriate technical documentation in a given error or maintenance situation in a factory. We call this application ProDok 4.0. We will demonstrate its use through two use cases: robotic application development and maintenance of industrial inspection machines. Although the use cases are very different, the underlying semantic application has a common software architecture.

## 12.2   Use Case 1: Robotics Application Development

State-of-the-art robots such as the KUKA LBR iiwa (Fig. 12.1) are designed to be deployed in a wide range of application domains. A key feature to enable this flexibility is the capability to sense forces and torques, and thus, to react to haptic user interaction as well as physical contact or collisions with its environment. The ability to measure external forces and torques allows for the development of force-controlled robot applications, where the teach-in of exact positions is no longer required. Instead, an application could, for example, move the end effector towards a surface until it detects physical contact and then move along the surface while constantly applying a certain force. This allows for the development of sensitive joining or peg-in-hole applications without knowing the exact position of the surface, the hole, the parts to be joined, etc., while achieving as high a precision as could possibly be realised using a position control mode only.

These new features cause an increase in complexity when developing robotic applications. This is because force/torque measurements and the corresponding conditional application control mechanisms need to be mastered in addition to a correct and more

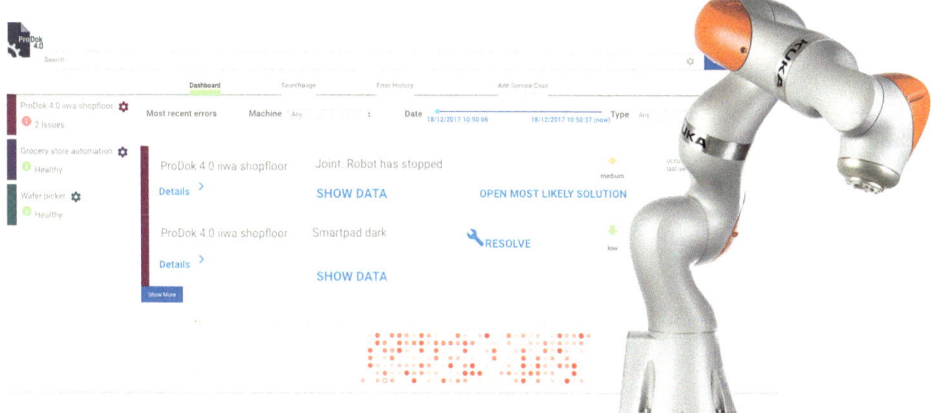

**Fig. 12.1**  LBR iiwa and ProDok 4.0 web interface

comprehensive configuration of the robot, end effector, workpiece, etc. Matching both the machine configuration and status as well as the developers knowledge level, our ProDok 4.0 application shall support the process of developing robotic applications by allowing access to helpful documentation for error situations with ease.

## 12.3    Use Case 2: Maintenance of Industrial Inspection Machines

In the production of glass, defects may occur. A glass inspection machine uses computer vision techniques for detecting such defects and for rating material quality [2]. The glass inspection machine consists of cameras and lights as well as servers with inspection software [12]. In Fig. 12.2, a typical glass inspection machine setup is shown.

An error may occur in every component of the setup, including the interconnections, and on both the software level and the hardware level. Errors in the inspection machine may lead to uninspected glass or glass of unknown quality and thus impact the plant yield. With today's quality requirements for glass products, only quality-inspected glass can be sold to customers.

Common inspection machine errors are camera failures, network errors, lighting issues, or incorrect configuration of external system parameters. The machine itself detects and reports known issues. Typically, the machine on its own cannot solve these issues, e.g. a lost communication signal with another node due to physical damage of the cable.

Less obvious or even previously unknown problems can cause an altered behaviour in the defects detection (over-detection or under-detection). Often, hints for these kinds of problems can be found provided that the distributed information the system generates at runtime is considered as a whole. Our approach aims at detecting errors within the inspection machine based on different indicators from this runtime information. After having identified an error, the system shall give a statement on its cause and, if possible, point to a solution to fix the problem.

**Fig. 12.2** Glass inspection machine setup

## 12.4   Requirements

Based on expert interviews conducted with personnel of manufacturers for smart factory equipment (robot application developers, inspection machine support engineers), the following requirements have been collected [1]:

R1) In case a machine error occurs in a smart factory, personnel shall be enabled to *resolve the error quickly and with little effort.*

R2) *Appropriate technical documentation* shall be provided, indicating causes and solutions of machine errors.

R3) The provided technical documentation shall *match the machine context* in which the error occurred.

R4) The technical documentation shall be provided *automatically*, alerting the user as soon as the machine error occurs.

R5) *Devices* shall be supported which support the smart factory workflow, e.g., desktop computer, tablet PC, smartphone, or smartwatch.

R6) *Usability* of the user interface shall be high.

R7) *User interaction* shall be *fast* so as not to disturb the personnel's workflow.

## 12.5   Architecture

### 12.5.1 Information Architecture

We explain the interaction concept with the robotics example use case previously outlined. When the robot stops during hand guiding, the user is alerted. This alert may be pushed to a suitable device, e.g., the development workstation, a tablet PC, or even a smartwatch. The alert indicates the symptom of the machine error, i.e. 'Joint: the robot has stopped'. See Fig. 12.3 for a screenshot of a dashboard view on the development workstation.

An important aspect of the screen design is the clarity of information presentation, following the ISO Standards 9241-110 [10] and 9241-210 [11]. The user-centred design

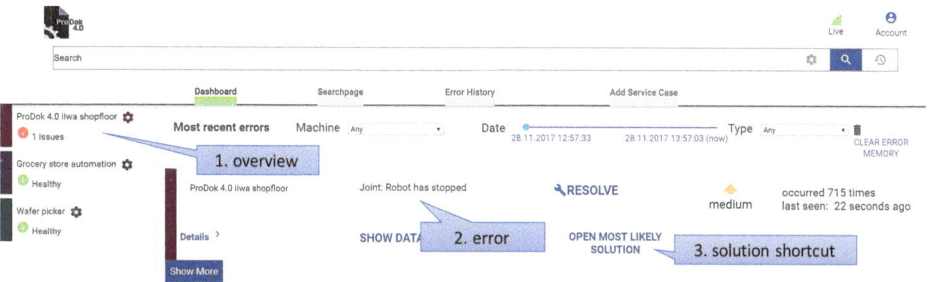

**Fig. 12.3**  Dashboard view. (Adapted from [1])

approach is applied, which includes close contact with the end user to define requirements collaboratively and test intermediate prototypes iteratively [6].

The dashboard component implements the interaction design pattern of sequence-of-use, which follows the mental model of spatial alignment of semantically related objects [14]. Elements that share a semantic relationship are grouped and the interaction design provides subsequent interaction steps for the main tasks.

The dashboard's most important components are:

(a) The overview functionality of all connected devices as a list (Fig. 12.3, Mark 1).
(b) A table containing the most recent errors for all connected devices. Each column displays the error symptom alongside a short cut towards the solution, e.g. 'Joint: the robot has stopped' (Fig. 12.3, Mark 2) and a corresponding navigation component reducing the effort a user has to invest for finding a solution ('open most likely solution') (Fig. 12.3, Mark 3).

With a single interaction, i.e., a click on 'open most likely solution' (Fig. 12.3, Mark 3), the user is provided with a solution to the most likely cause of this machine error. See Fig. 12.4.

The conceptual ideas of the Solution View (Fig. 12.4) are as follows:

(a) Present the solution to the machine error, e.g. the solution text 'Move joint out of maximum angle position' (Fig. 12.4, Mark 1).
(b) Provide a quick view to the user regarding error context and symptom, e.g. the blue headline 'LBR iiwa 14 R820 | Joint: the robot has stopped' (Fig. 12.4, Mark 2).

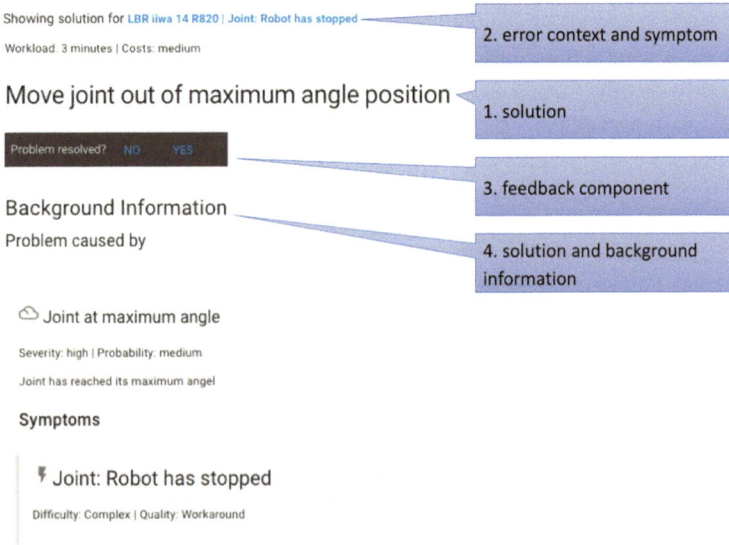

**Fig. 12.4** Solution view. (Adapted from [1])

(c) Collect user feedback for a solution, e.g. 'Problem resolved?' (Fig. 12.4, Mark 3).
(d) Display solution background information to the user, matching both error context and error, e.g. the cause and symptom (Fig. 12.4, Mark 4).
(e) In case several causes exist (here there is only a single cause identified: 'Joint at maximum angle'), the less likely solutions are sorted by likelihood descending, and displayed on the solution view following the most likely solution.

## 12.5.2 Ontology

An *ontology* specifies concepts and their relationships. One purpose of an ontology is to bridge terminology across different domains [3]. In this case, the event data and machine data, both originating from the machine, are bridged to the technical documentation. The ontology can be queried to retrieve *symptoms, causes,* and *solutions (SCS)*, matching both product and error data. See Fig. 12.5 for an example following the W3C recommendation for concepts and abstract syntax [17].

Within the ontology, different concepts are modelled:

(a) Hierarchies of products and errors, both interlinked, e.g.:
    'LBR iiwa R820' 'isA' 'LBR iiwa'
    'LBR iiwa' 'has Error' Joint: maximum angle reached, robot stopped'
(b) Technical documentation split into symptoms, their causes and solutions (SCS) with linkages to (a), e.g.:
    'Joint: maximum angle reached, robot stopped' 'hasSymptom' 'Joint: Robot has stopped'
    'Joint: Robot has stopped' 'hasCause' 'Joint at maximum angle'
    'Joint at maximum angle' 'hasSolution' 'Rotate joint'
    all linked to 'LBR iiwa' via 'hasProduct'

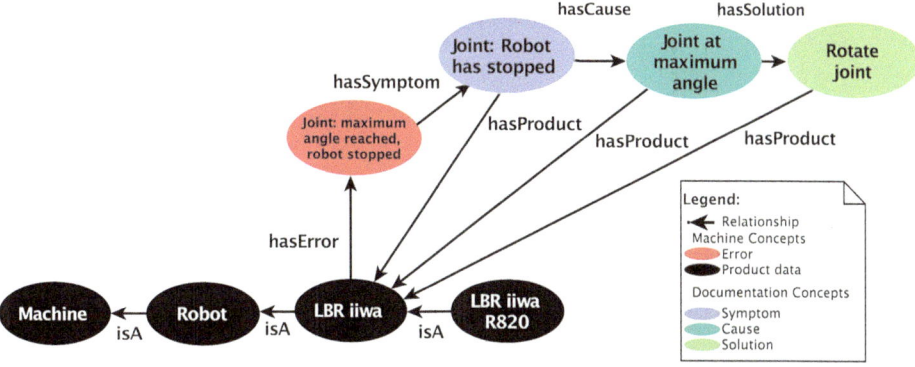

**Fig. 12.5** Ontology example [1]

The ontology enables modelling transitive relationships like 'isA'. 'LBR iiwa R820' has no direct relationship with any error, symptom, cause, or solution. However, due to the 'isA' relationship with 'LBR iiwa', the relationships to the respective error, symptom, cause, and solution can be inferred.

In addition the concepts and relationships shown in Fig. 12.5, may have additional attributes. For example, a solution may have an attribute containing a detailed description on how to apply the solution. An additional attribute for SCS may contain information about the target user group.

### 12.5.3 Software Architecture

The software architecture is shown below in Fig. 12.6 as an UML class diagram. It consists of three *layers* [16]: *Presentation, Logic,* and *Data.* Each layer encompasses different modules. Following Fig. 12.6, the purpose of each module is described.

**Presentation Layer:** Contains the Graphical User Interface (*GUI*) which serves as an entry point for the user.

**Logic Layer:** This layer encompasses two modules: (a) *Semantic Knowledge Retrieval* for providing accurately matching documentation and (b) *User Feedback Adapter* for handling user feedback.

**Data Layer:** Three modules are located here. (a) The *Machine* sends out event and context information, (b) the *Ontology* contains the hierarchies of products and errors, both interlinked, as well as the modularised technical documentation and (c) The User Feedback Store contains collected user feedback.

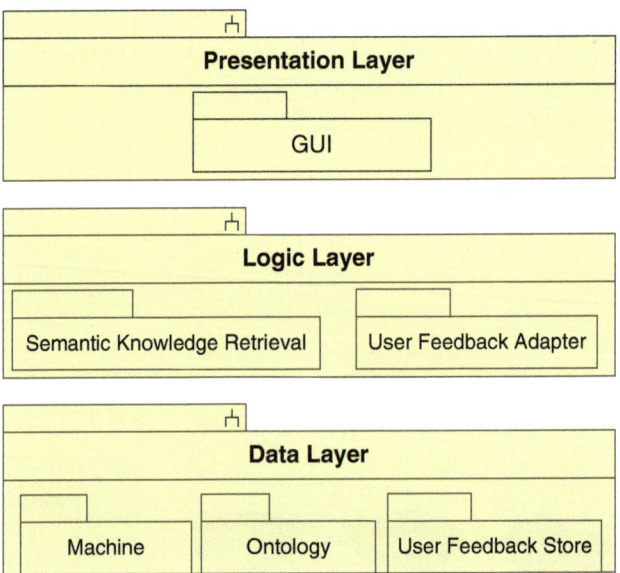

**Fig. 12.6** Software architecture [1]

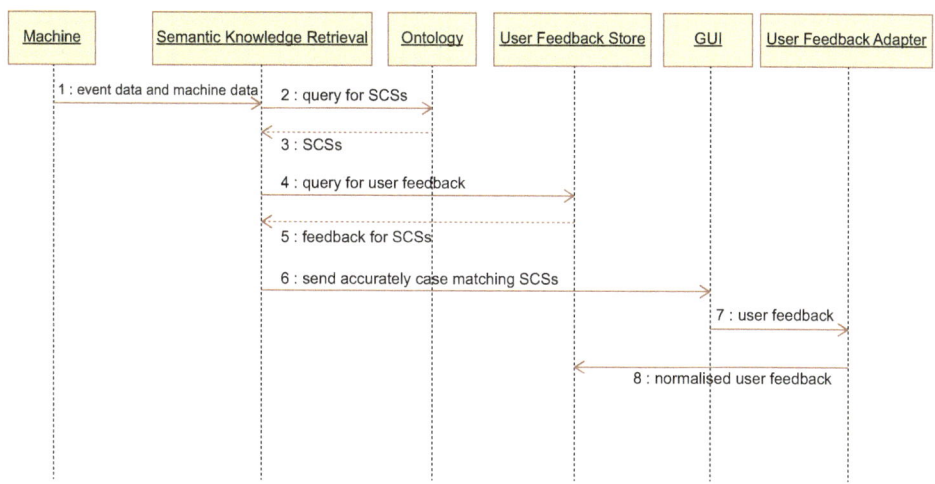

**Fig. 12.7** Communication between components [1]

Figure 12.7 provides an inside view of the communication between components as an UML sequence diagram.

Step 1: When an error occurs, the machine sends *event data* and *machine data* to the '*Semantic Knowledge Retrieval*' component. *Event data* describes the machine error in detail, e.g. 'Joint 3 maximum angle reached, robot stopped' on 13:45:17. *Machine data* contains context information, e.g. the robots' digital identification plate with manufacturer and type, e.g., 'KUKA' and 'LBR iiwa 14 R820'.

Steps 2,3: Modularised technical documentation is retrieved from the ontology by querying for the event data in combination with the machine data. The modularised technical documentation consists of the documentation fragments *symptom*, *cause*, and *solution (SCS)*.

Steps 4,5: User feedback is queried for each of the previously retrieved SCSs.

Step 6: Technical documentation accurately matching both the machine event and the personnel's preference is forwarded to the 'Graphical User Interface' (GUI) component.

Steps 7,8: When the GUI sends *user feedback*, it is normalised and saved in the feedback store.

## 12.6  From Raw Data to Semantic Context

In some situations, machine errors are less obvious than in the example above, where the machine sends an error message like, e.g., 'Joint 3 maximum angle reached, robot stopped'. For example, consider a communication failure between two components of the inspection machine. Such an error can only be detected by observing that regular messages have been missing for an unusual amount of time.

In such situations, raw data needs to be semantically enriched to get a semantic context. We call the process from raw data to semantic context the Semantic Fusion Process (SFP) [13]. Figure 12.8 shows the SFP as a BPMN diagram. In the following, the SFP is explained employing the use case of 'Maintenance of Industrial Inspection Machines' for glass production.

Any internal state changes, exceptions, sensor data, and communication between components (software or hardware components) inside the glass inspection machine is saved into log files. Each line in a log file corresponds to a log event (raw event). Raw events are

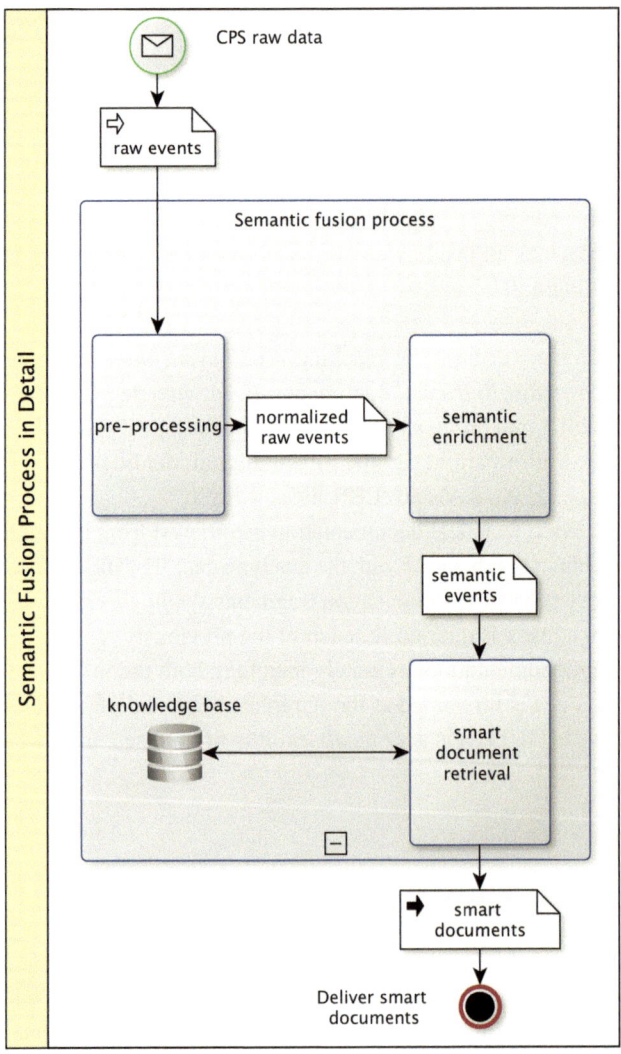

**Fig. 12.8** Semantic fusion process. (Adapted from [13])

the inbound data of the SFP. Multiple log events are collected within a *data stream* and delivered in real-time into the process.

The *Semantic Fusion Process* consists of three steps. In the *pre-processing* step, differently formatted log events are normalised to a defined schema. In the subsequent *semantic enrichment* step, normalised raw events are refined using analysis methods, generating semantic events. In the *smart document retrieval* step, these semantic events enable a fine-grained query to the knowledge base, thus leading to a composition of smart documents.

## 12.6.1 Pre-processing

In order to support multiple log events with different encodings and formats from different software and hardware modules, a pre-processing step is needed. Implementing the idea of [15] on the input data, log events are normalised using a defined schema, hence decoupling the SFP from the inbound data format. Figure 12.9 shows the pre-processing step with the example event *GlassBreakBegin*.

*GlassBreakBegin* indicates the detection of a break within the glass layer. It is reported by the glass inspection machine as a formatted log message as shown in Fig. 12.9. Each event has general attributes such as timestamp, context, and type. The 'timestamp' attribute indicates the time at which the error occurred. The 'context' attribute reflects the origin within the machine, e.g., for a camera event on slave 1 '/slave1/camera'. The 'type' attribute classifies the event, here *GlassBreakBegin*. In addition, an event may have specific event attributes, e.g., ticks (conveyor belt position). The formatted log message of the raw event gets parsed and the extracted data are stored in a normalized raw event object.

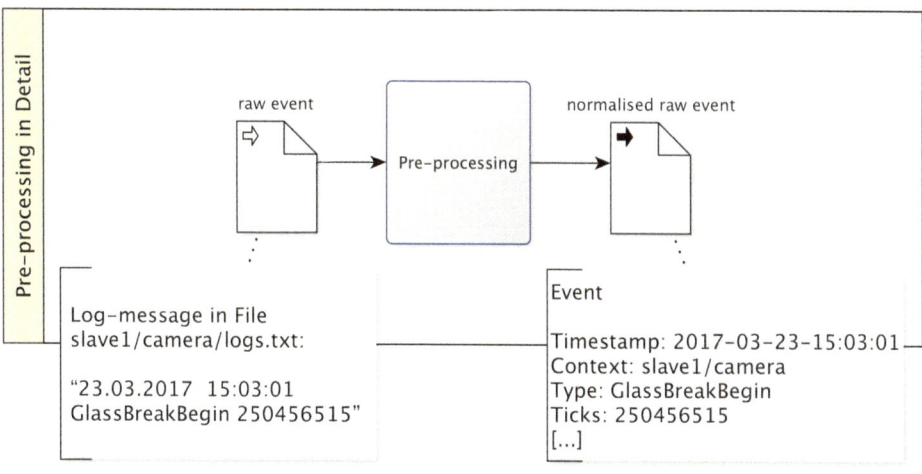

**Fig. 12.9** Pre-processing of the event GlassBreakBegin [13]

## 12.6.2 Semantic Enrichment

Each normalised raw event alone may not be sufficient to identify the machine problem. Consequently, in a second step, the normalised log events get lifted semantically in the semantic enrichment process. Figure 12.10 shows the semantic enrichment process with four subprocesses. Normalised raw events are input data for the semantic enrichment process. The process can apply *filters, pattern matching, value progression analysis, and time progression analysis* to generate semantic events.

In addition, semantic events can be used as input data for semantic enrichment. In the case of an inbound semantic event, higher semantic events can be generated.

Figure 12.11 displays a filtering operation on a normalised event stream, here filtering *GlassBreak* events.

On the left side, the unfiltered data stream is shown containing multiple normalised events. On the right side, only filtered events are shown. For the filter operation, we use pseudocode similar to common complex event processing (CEP) languages as used in CEP tools such as Apache Flink.

Figure 12.12 shows an example for pattern matching operation, identifying corresponding *GlassBreakBegin* and *GlassBreakEnd* events.

The pseudo code specifies a pattern in which a "*GlassBreakBegin*-event-immediately-followed-by-an-*GlassBreakEnd*-event" is detected in the data stream. Each pattern recognised can be used to generate a semantic *GlassBreakDetected* event.

Figure 12.13 shows an example of a value progression analysis for generating the semantic *SpeedChanged* event. The speed of the inspection process may vary, due to the versatility of the glass production process. A speed change may affect defect detection and, therefore, is important semantic information.

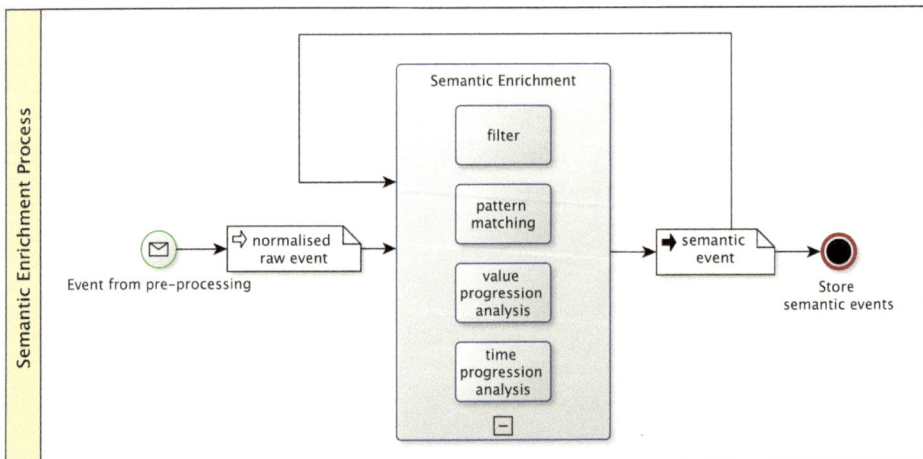

**Fig. 12.10** Semantic enrichment process [13]

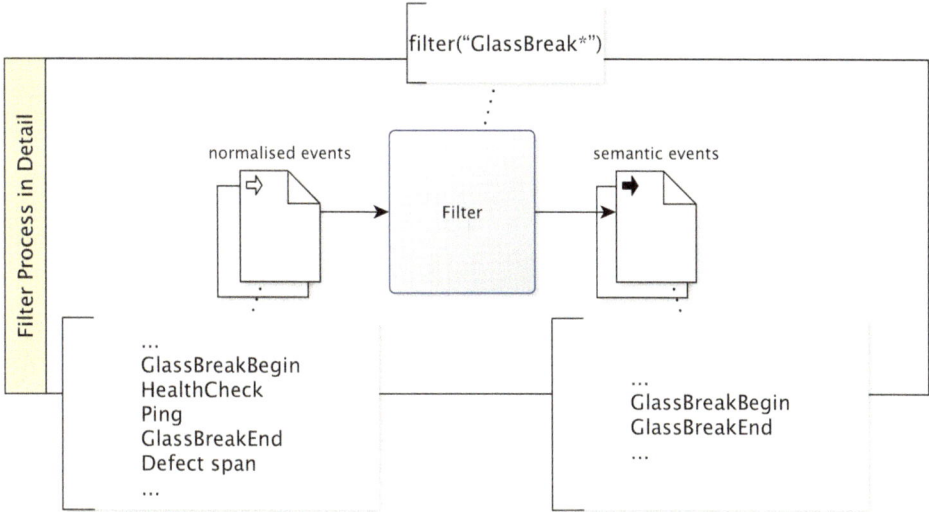

**Fig. 12.11** Example of a filter process [13]

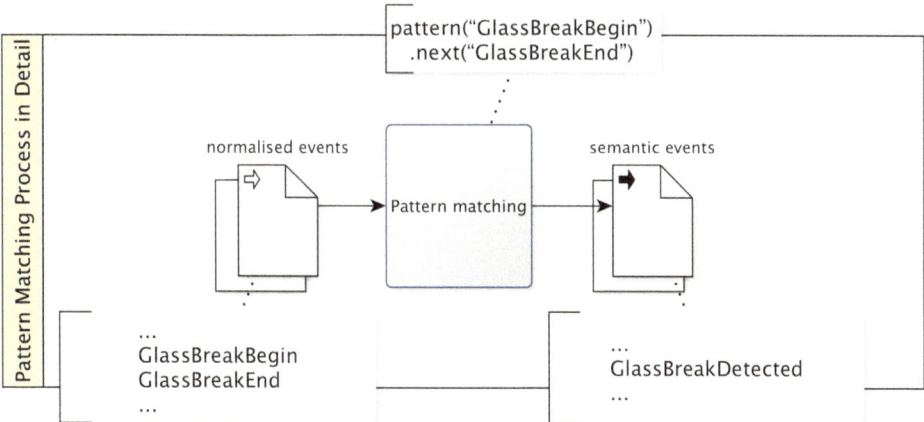

**Fig. 12.12** Example of a pattern matching process [13]

Each *SpeedCheck* event delivers a snapshot of the speed of the conveyor belt. The pseudo code for implementing the value progression analysis uses sliding time windows of size 5000s with an overlap of 10s. For each time window, it is checked whether the difference in speed exceeds a threshold (here 0.5s). In this case, a new semantic event *SpeedChanged* is generated.

Figure 12.14 illustrates the time progression analysis with the example of the *SignalLost* event. The *SignalLost* event indicates a connectivity failure, e.g., between a camera and a server.

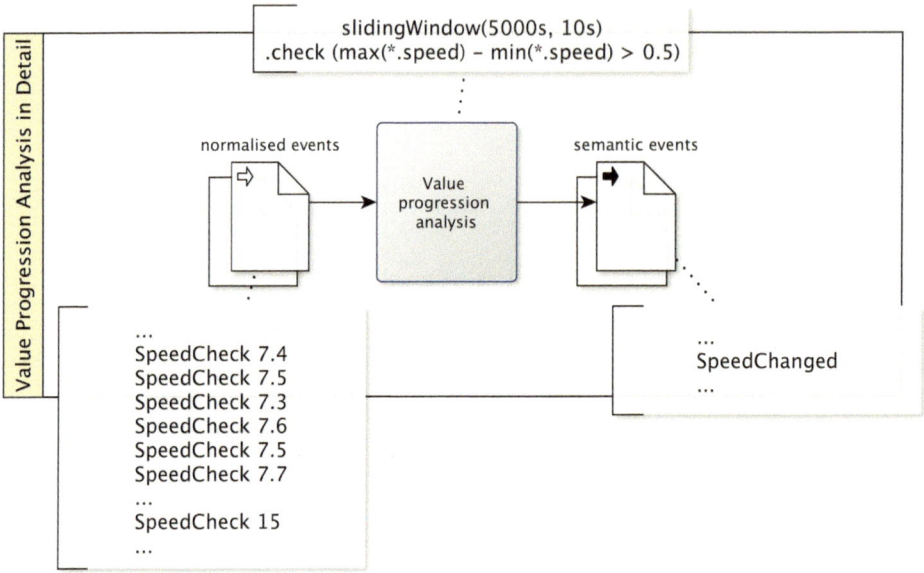

**Fig. 12.13** Example of a value progression analysis process [13]

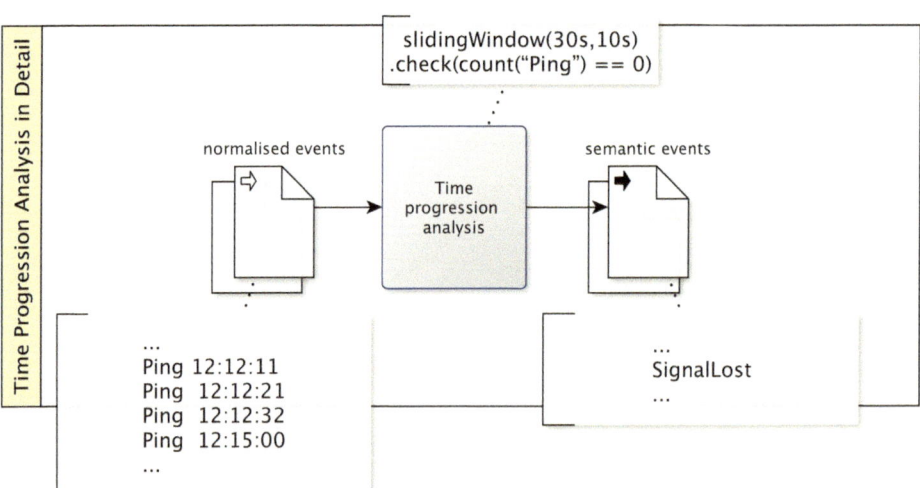

**Fig. 12.14** Example of a time progression analysis [13]

Each component within the glass inspection machine regularly sends ping events. If no ping event occurs within half a minute, it is considered as a loss of signal. The pseudocode shown in Fig. 12.14 uses a sliding time window of 30s with an overlap of 10s. If no ping event occurred within the time window, then a *SignalLost* event is generated.

As shown in Fig. 12.10, semantic events generated by semantic enrichment processes can be used as input for other semantic enrichment processes. So, a chain of successive enrichment processes can be established. For example, high conveyor belt speed typically implies a thinner glass layer. With the glass thickness, the properties of the defects change, and in the transition between different thicknesses many defects may occur. The semantic event *SpeedChanged* introduced above may then be used for generating a semantic *ThicknessChanged* event.

## 12.7    From Semantic Context to Appropriate Documentation

Technical documentation is stored in a *knowledge base*. Technical instructions are provided in the form of *smart documents*. A smart document is modularised, containing the building blocks symptom, cause, and solution. This structure forms the schema of the knowledge base. We call it the *symptom / cause /solution model (SCS model)*.

A *symptom* is a misbehaviour of any form as visual, physical or nonphysical (software-related) aspect [7]. A cause can be the origin of a symptom, but one cause can be linked to multiple symptoms and symptoms can have multiple causes. In addition, a solution covers one or more causes and a cause can be fixed by multiple solutions. On top of this, each solution can have a *semantic context* defining the scope of the solution, e.g., *server* or *camera* (see Fig. 12.15).

Through this modularised structure, it is possible to assemble a smart document consisting of a symptom, a cause, and solutions according to the semantic context for a given semantic event. The smart document is the final output of the SFP, semantically interlinking semantic events and documentation for a machine-specific problem.

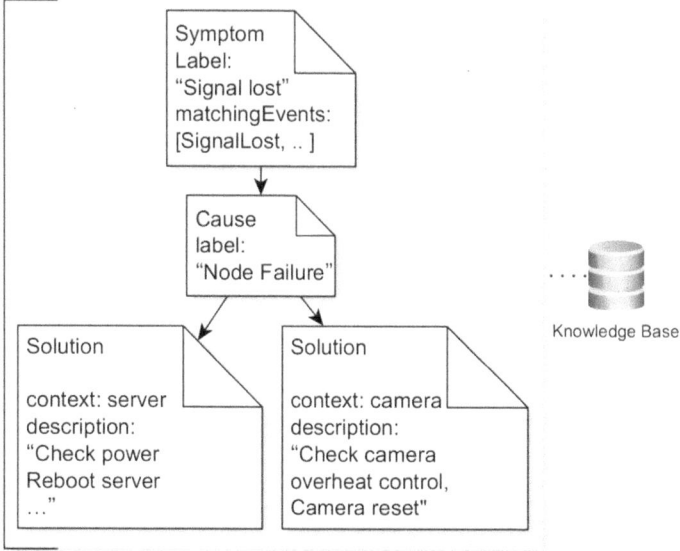

**Fig. 12.15**  SCS model with example data [13]

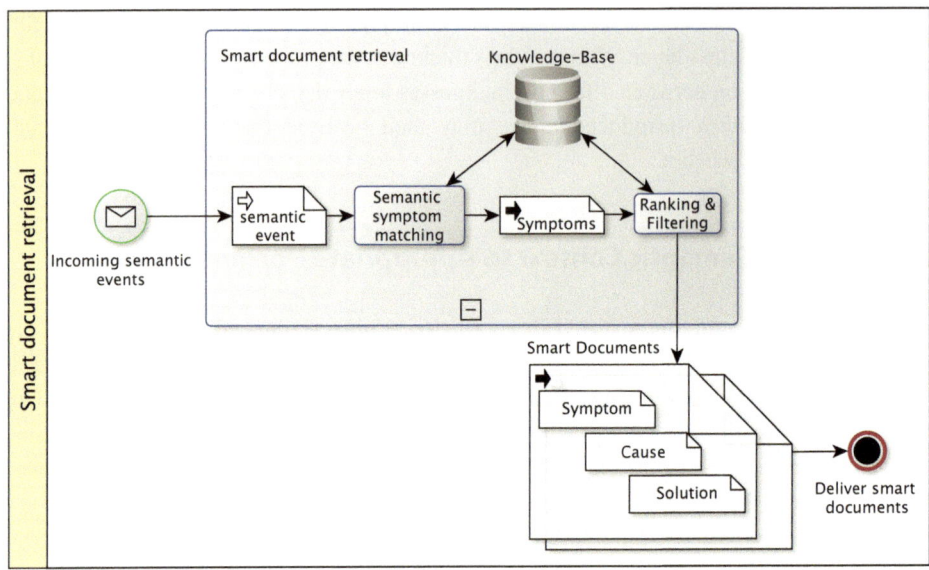

**Fig. 12.16** Smart document retrieval process [13]

For instance, the previously generated semantic event *SignalLost* can be mapped onto the symptom 'signal lost', and the 'signal lost' symptom within the knowledge base has a 'node failure' cause with two different solutions.

In Fig. 12.16, the smart document retrieval process is shown, consisting of a 'semantic symptom matching' process and a 'ranking and filtering' process.

The 'semantic symptom matching' process queries the knowledge base, e.g., querying on a 'matchingEvents' attribute to determine the right symptom. Multiple symptoms can have the same 'matchingEvent' like *SignalLost*. In order to provide suitable smart documents, the symptoms get filtered by semantic context. A ranking can be applied according to filtered causes and solutions. For example, the frequency of occurrence of a problem-cause-solution triplet in the past may be used as a ranking criterion.

## 12.8    Recommendations

We have successfully implemented an application for providing context-aware documentation in the smart factory. We summarize our main learnings from implementing and applying it to both use cases with the following recommendations:

1. Normalise raw data from machinery using a common data format and a (simple) ontology.
2. Queueing technology such as Apache Kafka is mature and scalable and well-suited for communicating events from machinery.

3. Complex Event Processing (CEP) technology such as Apache Flink is mature and scalable and well-suited for semantically enriching raw data from machinery. Functionality provided includes filtering, pattern matching, machine learning, value progression analysis, and time progression analysis.
4. Modular documentation is required when automatically matching technical documentation to machinery data. We recommend separating symptoms, causes, and solutions. The ontology links the machinery with their events and the events with their technical documentation.
5. Check for existing ontologies before developing a custom ontology. This includes product hierarchies and domain vocabularies, e.g. the eCl@ss product and service classification [4], IEEE Standard Ontologies for Robotics and Automation [8], IEEE Suggested Upper Merged Ontology (SUMO) [9], etc.
6. Emphasize the usability and user experience of the semantic application in order to increase acceptance by users.

## 12.9  Conclusion and Outlook

Documentation in the smart factory is a hot topic. In the project ProDok 4.0, we have regularly invited an open industry board to discuss the applicability of our solution in other corporate use cases. The enormous interest in the topic even surprised us. Many corporations, particularly in the manufacturing sector, face the constant challenge of finding appropriate documentation in error and maintenance situations. Our solution of semantically selecting appropriate documentation based on the current machine context has proven highly attractive to the industry board members.

Where is the road leading? Distilled from intense discussions, here are some suggestions for future work:

- *User-specific information delivery:* Context-awareness may also take into account the skill level and role of the user, providing even more appropriate documentation.
- *Documentation at the point of action:* This requires the use of mobile devices, or, more appropriately, augmented reality devices like Google Glass. Alternative user interaction mechanisms like voice control may be necessary in such settings.
- *Predictive maintenance*: Much better than fixing an error is avoiding it altogether. For certain situations, our solution can be extended to predict errors and maintenance situations based on common patterns and inform the user before issues occur.
- *Executable solutions*: the connectivity of machines in the smart factory allows not only to retrieve context for selecting appropriate documentation, but also to execute operations. Therefore, the user could be offered an additional option 'automatically apply solution in certain situations'.

Semantic applications can really make a difference.

## References

1. Beez U, Bock J, Deuschel T, Humm BG, Kaupp L, Schumann F (2017) Context-aware documentation in the smart factory. In: Proceedings of the collaborative European research conference (CERC 2017), Karlsruhe
2. Beyerer J, León FP, Frese C (2016) Automatische Sichtprüfung: Grundlagen, Methoden und Praxis der Bildgewinnung und Bildauswertung. Springer, Berlin
3. Busse J, Humm B, Lubbert C, Moelter F, Reibold A, Rewald M, Schluter V, Seiler B, Tegtmeier E, Zeh T (2015) Actually, what does "ontology" mean?: a term coined by philosophy in the light of different scientific disciplines. J Comput Inf Technol 23(1):29. https://doi.org/10.2498/cit.1002508
4. eCl@ss (2017) Introduction to the eCl@ss standard [online]. Available at: https://www.eclass.eu/en/standard/introduction.html. Accessed 21 Sept 2017
5. EPC (2006) Directive 2006/42/EC of the European Parliament and of the Council of 17 May 2006 on machinery, amending Directive 95/16/EC (recast). Off J Eur Union [online]. Available at: http://eur-lex.europa.eu/legal-content/EN/TXT/?uri=uriserv:OJ.L_.2006.157.01.0024.01.ENG&toc=OJ:L:2006:157:TOC. Accessed 21 Sept 2017
6. Garrett JJ (2011) The elements of user experience: user-centered design for the web and beyond, 2nd edn. New Riders, Berkeley
7. Hornung R, Urbanek H, Klodmann J, Osendorfer C, van der Smagt P (2014) Model-free robot anomaly detection. In: 2014 IEEE/RSJ International conference on intelligent robots and systems, pp 3676–3683. https://doi.org/10.1109/IROS.2014.6943078
8. IEEE (2015) 1872-2015 IEEE Standard ontologies for robotics and automation. IEEE Robot Autom Society [Online]. Availabe at https://standards.ieee.org/findstds/standard/1872-2015.html. Accessed 25 Sept 2017
9. IEEE, Adam Pease (2017) Suggested upper merged ontology (SUMO) [Online]. Available at http://www.adampease.org/OP/. Accessed 25 Sept 2017
10. ISO (2006) DIN ISO 9241-110:2006: 'DIN EN ISO 9241-110 Ergonomie der Mensch-System-Interaktion – Teil 110: Grundsätze der Dialoggestaltung (ISO 9241-110:2006)'. Deutsche Fassung EN ISO 9241-110:2006: Perinorm [Online]. Available at http://perinorm-s.redi-bw.de/volltexte/CD21DE04/1464024/1464024.pdf. Accessed 28 June 2013
11. ISO (2010) DIN ISO 9241-210:2010: 'Ergonomie der Mensch-System-Interaktion – Teil 210: Prozess zur Gestaltung gebrauchstauglicher interaktiver Systeme' [Online]. Available at http://perinorm-s.redi-bw.de/volltexte/CD21DE05/1728173/1728173.pdf. Accessed 29 Oct 2014
12. ISRA VISION AG (2015) The NEW Standard In Float Glass Inspection FLOATSCAN-5D Product Line, ISRA VISION AG [online]. Available at: http://www.isravision.com/media/public/prospekte2013/Brochure_Floatscan_5D_Product_Line_2013-05_EN_low.pdf. Accessed 13 June 2017
13. Kaupp L, Beez U, Humm BG, Hülsmann J (2017) From raw data to smart documentation: introducing a semantic fusion process. In: Proceedings of the collaborative European research conference (CERC 2017), Karlsruhe
14. Koffka K (2014) Principles of gestalt psychology. Mimesis Edizioni, Milan
15. Nuñez DL, Borsato M (2017) An Ontology-based model for prognostics and health management of machines. J Ind Inform Integr [online]. Available at: http://www.sciencedirect.com/science/article/pii/S2452414X16300814?via%3Dihub. Accessed 21 Sept 2017
16. Starke G (2015) Effektive Software-Architekturen: Ein praktischer Leitfaden, 7th edn. Hanser, München. https://doi.org/10.3139/9783446444065
17. W3C (2014) RDF: '1.1 Concepts and Abstract Syntax' [Online]. Available at https://www.w3.org/TR/rdf11-concepts/. Accessed 16 June 2017

# Knowledge-Based Production Planning for Industry 4.0

# 13

Benjamin Gernhardt, Tobias Vogel, and Matthias Hemmje

**Key Statements**

1. The production planning of a new product's manufacturing process nowadays takes place in various partial steps and these planning steps are mostly executed in different locations, executed by different companies, potentially distributed all over the world, and this process has to be able to adapt quickly to changing circumstances.
2. When different companies work together on a joint production planning, they exchange, e.g., component production planning, i.e., manufacturing information between different distributed (planning) subsystems.
3. Therefore, a state of the art production planning must be dynamic, fast, decentralized, and always be available from everywhere.
4. Therefore, the joint planning of a product requires a cloud-based, collaborative, co-creative and adaptive production process planning approach.
5. Collaborative Adaptive (Production) Process Planning can be supported by semantic approaches for knowledge representation and management, as well as knowledge sharing, access, and reuse in a flexible and efficient way.
6. Semantic representations of such knowledge integrated into a machine-readable process formalization is a key enabling factor for sharing such knowledge in cloud-based knowledge repositories.

B. Gernhardt (✉) · T. Vogel · M. Hemmje
University of Hagen, Hagen, Germany
e-mail: benjamin.gernhardt@fernuni-hagen.de; tobias.vogel@fernuni-hagen.de; matthias.hemmje@fernuni-hagen.de

© Springer-Verlag GmbH Germany, part of Springer Nature 2018
T. Hoppe et al. (eds.), *Semantic Applications*,
https://doi.org/10.1007/978-3-662-55433-3_13

## 13.1   Introduction and Motivation

Today and tomorrow – in the era of **Digital Production Environments** and **Industry 4.0** – the production planning and manufacturing of a new product takes place in various partial steps, mostly in different locations, potentially distributed all over the world. In this application context, ***Collaborative Adaptive (Production) Process Planning (CAPP)*** [2] can be supported by semantic product data management approaches enabling production-knowledge representation and management as well as knowledge sharing, access, and reuse in a flexible and efficient way. To support CAPP application scenarios, semantic representations of such production-knowledge integrated into a machine-readable process formalization is a key enabling factor for sharing such explicit knowledge resources in cloud-based knowledge repositories. We will now introduce such a method supporting semantic representations of production-knowledge integrated into a machine-readable process formalization and a corresponding prototypical Proof-of-Concept (PoC) implementation called ***Knowledge-Based Production Planning (KPP)***.

When, e.g., ***Small and Medium Enterprises (SMEs)***, work together on production planning for a joint product, they exchange component production planning and manufacturing information between different distributed planning subsystems. This planning can happen within one organization or even across organizational and production domain boundaries. Also, the use case that ***Original Equipment Manufacturers (OEM)*** works with several global SME partners and suppliers is not unusual. The CAPP-4-SMEs project [2] has explicitly defined and achieved the goal to research and support the field of CAPP in the area of SMEs, e.g., to achieve a more optimized collaborative manufacturing value chain [3].

Typically, such production planning information is exchanged by means of applying the already well-established ***Standard for the Exchange of Product model data (STEP)*** [4]. In order to obtain a computer-interpretable production planning knowledge representation that goes beyond current ***STEP expressiveness***, a machine readable semantic web based knowledge representation is a key enabling factor. At the same time, such a semantic web based representation enables the storage, management, and sharing of such knowledge in cloud-based knowledge repositories and can assist the preparation of planning processes. ***Knowledge-based Process-oriented Innovation Management (German:Wissensbasiertes Prozessorientiertes Innovations Management (WPIM))*** [2, 5], provides the underlying basic semantic web methods for semantic innovation process representation, annotation, and management to our KPP method.

Furthermore, so-called ***Function Block (FB) Domain Models*** serve as a high-level knowledge resource templates for planning processes and are based on established engineering knowledge representation models. Thus, collaborative planning and optimization are made possible for mass production or for recurring routine tasks in a machine-readable and integrated presentation. In this way, knowledge can be shared in distributed semantic knowledge repositories in order to cross-/inter-link processes collaboratively, for example, to reproduce or to annotate them.

Wang et al. have introduced a method for representing web-based *Distributed Process Planning (DPP)* activities in [3], [6], and [7], parallel to the development of the WPIM method. The necessary concepts and terms of the DPP method will be introduced in here, based on slightly adapted excerpts from [6]. We have already described the necessary implementation and application of the DPP approach and its semantic web integration for KPP within a so-called *Mediator Architecture (MA)* in [1]. Such MAs are solving semantic integration challenges and integrating several local knowledge sources into a global semantic repository.

In addition, KPP is a Proof-of-Concept implementation of the ISO/DIS 18828-2 standard [8]. As mentioned, KPP is based on a semantic architecture that supports a planning process step by step, from the identification and preparation of planning raw data to a finished machine-readable and executable program code.

Finally, the ProSTEP iViP Association [9] recently published a White Paper, called "Modern Production Planning Processes" [10] that is also based on the currently emerging ISO/DIS 18828-2 Standard [8]. This recommendation represents a formal end-to-end reference process that can be adapted to individual needs, the so-called *Reference Planning Process (RPP)*.

In the remainder of this chapter, we will explain KPP in detail. Further, as a basis for evaluation and validation, we use the KPP approach as a possible reference implementation of RPP. We also will demonstrate the usability and interoperability of the PoC implementation of KPP. This includes an integrated visually direct manipulative process editor. Moreover, we will illustrate the first prototype of the KPP MA including a user-friendly query library based on the KPP ontology.

## 13.2   Knowledge-Based Production Planning

KPP as a reference implementation of the formal concept of the RPP recommendation requires a brief explanation. RPP is an end-to-end reference process that can be adapted to individual needs. It consists of three maturity level-related phases: **Concept Planning**, **Rough Planning**, and **Detailed Planning**. Thus, it can be seen as a high-level template for creating a concrete production planning process that takes individual company-specific and location-specific conditions into consideration. KPP takes advantage of this concept and integrates all its advantages into one integrated, distributed, and collaborative three-level approach (displayed in Fig. 13.1) for supporting production planning. Furthermore, KPP does this in a knowledge-based way by integrating production planning knowledge resources along process representations of the planning process.

Hence, KPP enables the mapping of a **Supervisory Plan** onto an **Execution Control Plan** and this onto an **Operational Plan** in an optimized manner. In the understanding of WPIM which is underlying KPP, the DPP planning process and resource knowledge are represented by planning activities consuming and producing planning knowledge resources.

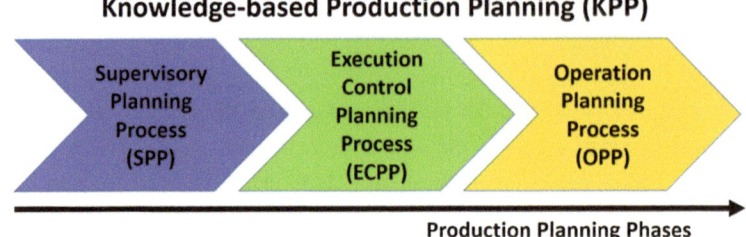

**Fig. 13.1** KPP process phases

These activities can be FBs over all levels of CAPP activities from *SP Process (SPP)* activities through *ECP Process (ECPP)* activities to *OP Process (OPP)* activities. A more detailed description can be found in the related paper [11].

The corresponding KPP mediation process is also performed in a three-level MA. Figure 13.2 displays the three-level MA. The first mediator is called the **SPP Mediator** and integrates MFBs and other relevant and potentially distributed resources for the SPP activity. A downstream DPP mediation can be implemented by two analogously derived additional mediators on the second and the third DPP level. On the second level of the MA follows the deduced and so-called **ECPP Mediator** which supports the above-mentioned ECPP activity. They assimilated at least an earlier iteration of the SPP Mediator as MFB and an OFB of the subsequent **OPP Mediator** (level 3) and various other relevant and potentially distributed knowledge resources. Coming from the machining-data point of view, the corresponding up-stream mediation process starts from machines with a defined need of steering information, which can be harmonized by using wrappers and offering a mediated interface to clients. The third and final level of the MA of the KPP process forms the derived **OPP Mediator** and completes the mediation process. This integrates relevant and potentially distributed manufacturing knowledge resources as FBs and by the second level generated EFBs (ECPP Mediator) for the OPP activity.

This three-level architecture can support a collaborative distributed production planning knowledge management and information process, by providing knowledge resources and related data from distributed data repositories, combining various data models, schemata and corresponding formats into a single semantic-enabled global schema and format. Moreover, it enables mediation process requesting, accessing, and collecting/gathering/combining data from different distributed manufacturing- and planning knowledge resources.

In the following, we will briefly summarize the state of the art of all technologies and methods used in KPP. This includes function blocks as well as DPP modelling. Also, the necessary concepts of information integration, mediation, the concepts of the mediator architectures, and the semantic representation of WPIM will be introduced. This section is based on an excerpt from [6] and the WPIM-based semantic process-modelling.

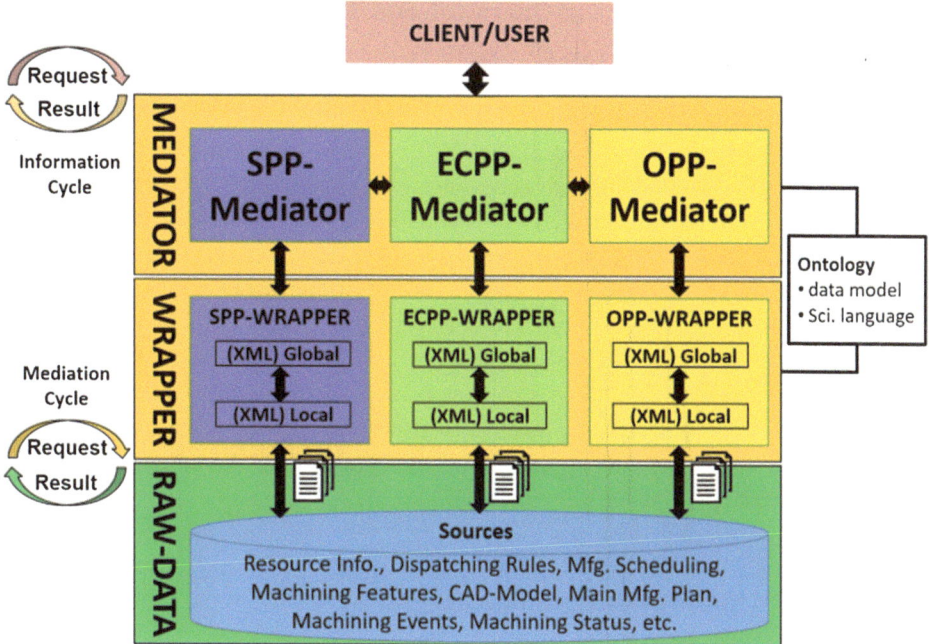

**Fig. 13.2** Conceptual architecture of the KPP mediator

### 13.2.1 Function Blocks

In the IEC 61499 standard [12], FBs were initially defined, and their development, implementation, and usage, including FBs in distributed process-, control-, and measurement systems in a component-oriented approach was explained [13]. This standard was jointly developed from the existing concepts of FB diagrams in the **Programmable Logic Controllers** language standard IEC 61131-3 [14] and standardization work concerning Fieldbus [14]. FBs are triggered by events and contain algorithms and an **Execution Control Chart** with input and output of data and events. An FB-related literature review to the research targeting the areas of machining and assembly is available in [6, 7].

### 13.2.2 Distributed Process Planning

The three planning processes of *Supervisory Planning (SP)*, *Operation Planning (OP)*, and *Execution Control Planning (ECP)* are the core components of DPP and are outlined in more detail in [6]. These processes are explicitly modelled in a conceptual *ICAM Definition for Function Modelling (IDEF0*, where *ICAM* is an acronym for *Integrated Computer Aided Manufacturing*) process formalization model together with their inter-relationship and data flow.

### 13.2.3 Meta, Execution, and Operation Function Blocks

MFBs are used to encapsulate machining sequences (machining features and setups) in this approach. They only contain general information about process planning of a product and act as high-level process templates. MFBs contain, e.g., tool path patterns and suggested cutting tool types, for subsequent manufacturing tasks.

Basically, an EFB can be created by instantiating a series of MFBs associated with a task that ready to be downloaded to a specific machine. Each manufacturing task matches its own set of EFBs so that the monitoring functions can be performed for each task unit.

OFBs and EFBs have the same structure. However, an OFB completes and specifies an EFB with more detailed, machine-specific data. This includes machining processes and operation sequences. Moreover, the actual values of variables in EFBs can override and be updated by the operation planning module, to make it locally optimized and adaptable to various events that happen during machining operations. In [6], Wang et al. use the two different terms of EFB and OFB.

### 13.2.4 WPIM in the Domain of Process Planning

To support capturing and usage of knowledge around innovation processes, the concept of WPIM [2, 5, 15] was developed. WPIM assumes that innovations have a knowledge and process perspective that need to be used in a combined manner. Therefore, activities of a process can be annotated with resources, such as experts and documents [15]. The WPIM application and its tool suite are based on the *Resource Description Framework (RDF)* [16] and enables semantic-based searching by using the *SPARQL Protocol And RDF Query Language (SPARQL)* [17]. Furthermore, the *Web Ontology Language (OWL)* [18, 19] allow the modeling of concepts in classes and replaceable relationships. These technologies provide a formal semantic representation and support the formal description of machine-readable knowledge.

WPIM offers the formal concepts *Master Processes (MP)* and *Process Instances (PI)* (see Fig. 13.3) as well as **Activities** and **Tasks**. Moreover, the separation of modelling and capturing generic and instance specific process knowledge is supported. By this means, the WPIM toolbox allows, in a seamless way, the reuse of process components and their associated machine-readable knowledge resources. To represent PLM data in the field of technical products and production processes, WPIM has already been applied for representation and modelling as well as to support executing processes and also planning processes. WPIM offers semantics that have the advantage of being easily exchangeable and machine-readable. This helps, for example, to plan cross-organizational and distributed productions.

To represent such processes in WPIM users can select classes of process components and resources. Furthermore, they can use the WPIM ontology repository to register an instance of a process, a process resource as well as a process component. To achieve this, users can select the process instance-, the component-, or the resource classification system

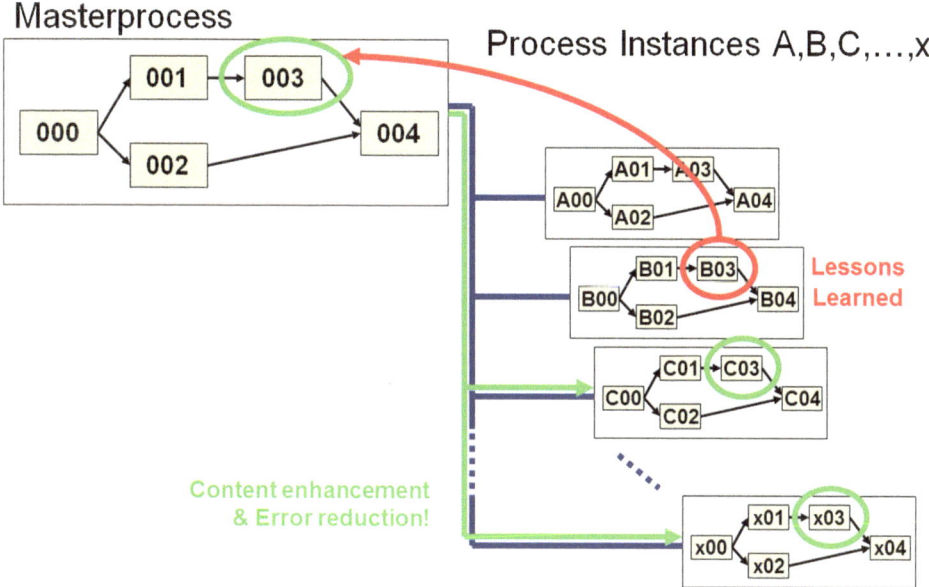

**Fig. 13.3**  Master process and process instances [5]

to be used as the global set of ontologies into which the knowledge contents and resource structure are to be mapped. In the next step, users can choose attributes for each resource class for populating virtual objects with content resources. This implies that users have to map the attributes of the resources to specific ontologies. For example, the mapping of a resource attribute onto an expert ontology. Finally, users can populate the resource instances and their specific content manually or pick the populating methods. In this way, users map attributes to classes in the ontology semi-automatically or manually. This works by using word-matching or other provided techniques, e.g., map "hole" from a product property ontology concept to the "drilled hole" concept in an ontology of machining features.

Nevertheless, the local data schemas of such a source have first to be registered before mappings can be established. Figure 13.3 displays the activity-based schemata of the implementation for representing the MP and PI resources. That means, during that execution of a first instance, the Lessons Learned can be stored within the MP (higher-level). Thus, this gathered information can be provided for the following executions of the process within the next PI (Fig. 13.3).

### 13.2.5  Master Processes and Instances of Processes, Activities and Tasks

**MPs** are generic high-level descriptions of processes. From a data set point of view, an MP in WPIM describes a data structure and attributes of a higher-level template. The semantic representation approach describes process structures and their attributes by using semantic representations, but goes beyond the sole representation of the process structural schema. WPIM offers semantic descriptions of MPs. This semantic MP schema exists as a formal generic description of a process and is thus independent of generated data instances during a certain execution of the process resources during execution of the entire process. In case of the CAPP adaptation, this could be production machines and production activities or documents and experts.

If a **Process** will be executed, data is gathered. From the data set point of view, WPIM describes this data as a PI. The structure of an **Activity** in WPIM is displayed in Fig. 13.4 and is used to store all incoming and outgoing data as well as states of activities. Moreover, PIs are well-ordered in a chronological way.

An **Activity** in WPIM needs well-defined input to generate a required output and contains one to many tasks. An instance of an Activity defines a cluster of tasks and thus can bundle tasks that are assigned to a single resource. For example, this kind of assignment can map tasks to a resource like a machine. However, this is just used in order to represent the execution of a machine operation or planning tasks which need to be executed by an expert, e.g., a planner.

A **Task** structure in WPIM, cannot be further split into subtasks. It is a simple action. Thus, a semantic data representation to values and the status when performing a task will be offered. For example, WPIM allows delegation of a task instance to various executing entities. Hence, to describe it in the context of planning tasks, a plan must be finalized by signing the plan and setting it into action. Obviously, to release a plan by a signature is a unique task and this signed task cannot be split. Therefore, either the plan is released by signature or it is not signed and thus not released.

An Activity consists of at least one or more Tasks (as displayed in Fig. 13.4) and represents the transformation of an input into an output of an Activity.

### 13.2.6 Semantic Integration Within Knowledge-Based Information Architectures

In the following, we will describe, based on a slightly adapted excerpt from [20], that data, information, and knowledge integration can be understood at varying levels of heterogeneity and interoperability. Nevertheless, several technical challenges must be overcome when trying to share distributed and heterogeneous data. Consider the example of two

**Fig. 13.4** Visualization of an activity as a set of tasks

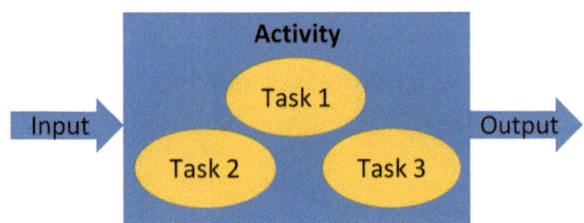

systems with data sets have to be made interoperable. Standards and technologies can be used to facilitate interoperability at different levels and to overcome the various kinds of heterogeneities. To achieve this, on the systems level, different operating systems, e.g., Windows, Linux MacOS, etc. or different data transport protocols, e.g., FTP, HTTP, etc., higher-level protocols for interoperation and discovery of web services may be found. There are also various application-specific system level issues besides from the generic issues of data access, transport, and remote execution, e.g., the choice and architecture of the mapping technology for the integration and mediation of knowledge and information resources. Furthermore, at the syntactic level, heterogeneities such as different data file formats independent of the type of content or knowledge resource and the corresponding representation format of information and knowledge must be considered.

### 13.2.7 Mediator Architectures

Originally introduced by Wiederhold in [1] in 1991, mediators are a standard approach in the construction of information system architectures. At that time, the semantic web did not yet exist, and the web was still in its infancy. Since then, the use and application of these architectures has grown into a de-facto standard for building web-based information systems supporting data, information, and knowledge integration. To provide unified data access to distributed heterogeneous data sets, database mediator systems can be used and thus overcome many of the interoperability challenges. A typical mediator architecture is depicted in Fig. 13.5. Several local data sources are "wrapped" (e.g. as XML [21] sources) and then combined into an integrated global view. Through this global view, the end user or a client application is provided with the illusion of querying one single, integrated database with one schema.

Thus, mediators serve to simplify, combine, integrate, reduce, and explain data. Furthermore, they are mainly used for providing a common access level onto different distributed data sources. The source wrappers not only provide uniform syntax, but also reconcile system aspects e.g., by a unified data access and query protocol (see e.g. [20, 22, 23]).

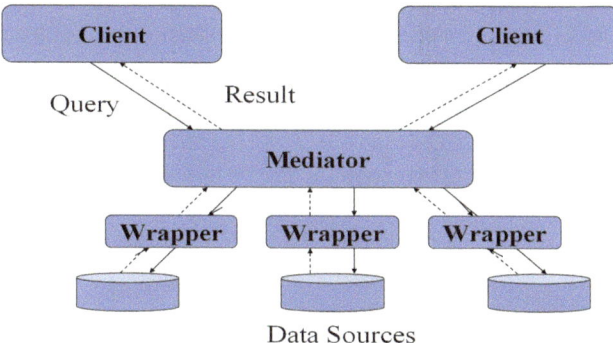

**Fig. 13.5**  Mediator architecture integrating data sources

In a conventional relational mediator system (based on XML, RDF [16], or OWL [24]), interoperability is facilitated at the structural level. Corresponding schema transformations will overcome differences in the schema as part of the view definitions for the global view. Therefore, terminological differences or other semantic differences are not adequately handled at the purely structural level (e.g. XML). For this reason, source schema and contents can be registered to an ontology represented in RDF or OWL. In this way, such an ontology encodes additional "knowledge" about the registered concepts. In the next section, we will explain how the system can evaluate high-level queries based on concepts through "ontology-enabling". However, these concepts are not directly stored in the source databases but indirectly linked via an ontology. It is the task of a mediator, to transform queries to the global schema into queries to the local source schemata and to collect the results as well as to integrate and link them. The global scheme is based on a suitable data model. For example, XML or RDF can be used for the representation. However, wrappers are software components that represent the contents of a data source for unification in another data model or schema. For example, XML wrappers are used to enable access to relational databases. The wrapper, specifically the coupling between source and mediator via the wrapper, allows the mediator the unified data access to the sources. This unified data access will be realized by creating a mapping between the data model of the mediator and the data model of the local source. At the same time, incoming requests can be translated into requests to the local source system of the mediator.

### 13.2.8 Ontologies in Information Integration and Mediation

Ontologies can be used in information integration systems that are based on a mediator architecture, to provide information at the level of conceptual models and terminologies. In this way, they facilitate conceptual-level queries against sources and resolve some of the semantic level heterogeneities between them. To translate queries from the global ontology to the local schema, wrappers use the mappings between the data source and ontology. This will also be used to translate content from the local schema to the global ontology. When required, the system can automatically use the subclass relation to expand concept queries. It should be noted that all system-registered ontologies can be considered as conceptual-level query mechanisms. The system can suggest suitable ontologies. The suggestion is based on first decision, "the user's choice of resources" and second decision, "the sources' schema information". Mediator systems, in special database mediator systems, can be used to provide unified data access to distributed and heterogeneous datasets [20].

As outlined in our previous paper [11], WPIM-based semantic representations of production knowledge can be applied to support DPP activities by a three-level KPP mediator architecture. In this way, the KPP mediator integrates distributed information and knowledge resources and resolves potential heterogeneity conflicts amongst them on all possible levels of heterogeneity (model, Schema and instance levels). This means that KPP enables an integration of, e.g., knowledge resources about product features, product design, machine/tool descriptions, machine/tool features, and process constraints in order to support collaborative information and planning processes that create an executable plan for a product production task.

## 13.3 Proof-of-Concept Prototype Implementation of KPP

The whole portfolio of different types of product data related to supporting KPP activities for component production is represented in XML language format to eliminate the communication barrier between different software, hardware, and specialized tools. For this purpose, the MA in KPP uses a wrapper technology with all common machine information and -codes encoded in XML by using, e.g., the STEP standard as a domain model. Initially, with a corresponding sample data set from the field of manufacturing, the SPP Mediator is utilized to demonstrate the functionality of a wrapper for **Product Design Information** and **Product Feature Information**. In the next step, the access to a typical **Main Manufacturing Plan** (see Fig. 13.6) represented as MFB from a local relational database source is implemented by an appropriate wrapper. It is integrated and represented in the global representation schema as STEP-compliant XML code. Through the normalization via the global schema, the parent mediator can get access to manufacturing

```xml
<?xml version="1.0" encoding="utf-8"?>
<job name="Job 122612" description="manufacturing job request" material="P" owner="Yuqian Lu" id="122612">
  <serviceRequirements>
    <deliveryTime>none</deliveryTime>
    <costExpectation>none</costExpectation>
  </serviceRequirements>
  <machiningFeatures>
    <machiningFeature featureType="General Open Pocket" externalId="" unit="Metric" name="General Open Pocket">
      <parameters>
        <parameter name="Dp" description="Depth of pocket">20</parameter>
        <parameter name="Rm" description="Minimum radius in concave corner">3</parameter>
        <parameter name="Wo" description="Width of open area">12</parameter>
        <parameter name="Cs" description="Corner style">Corner Break</parameter>
        <parameter name="Cb" description="Corner break">1.6</parameter>
        <parameter name="Rg" description="Floor radius">0</parameter>
        <parameter name="Wc" description="Width of 45° chamfer">0</parameter>
        <parameter name="Aw" description="Angle of wall">90</parameter>
        <parameter name="Wi" description="Smallest width of the gap">0</parameter>
        <parameter name="Vp" description="Volume of the pocket">0</parameter>
        <parameter name="Qw" description="Quality of wall surface">N10</parameter>
        <parameter name="Unit" description="Unit of Tool">None</parameter>
        <parameter name="FType" description="Solid/Indexable Tool">None</parameter>
        <parameter name="SBType" description="Shank/Bore Type">None</parameter>
      </parameters>
      <toolAssemblies>
        <toolAssembly referenceId="" />
      </toolAssemblies>
      <designModels>
        <model format="STEP">.\Models\PocketModel.stp</model>
      </designModels>
    </machiningFeature>
  </machiningFeatures>
</job>
```

**Fig. 13.6** Typical production data in XML format [25]

information and may offer it to the client and vice versa. Furthermore, the following code sample (see Fig. 13.6) from Y. Lu et al. [25] clarifies the representation of machining information that is integrated with the production plan information in the next step of the integration. This illustrated request represents the machining feature type, typical important parameters, and requirements which are important for the production process.

All remaining product information, manufacturing steps, requirements, and the CAD-Model, which are necessary to produce a product are now processed in a similar form in the SPP without being based on a concrete machine or tool. For example, SPP includes the "feature-based design", "fixture information" and "machining technology and constraints" of a product. Now the mediator specifically the wrapper has to be implemented and allows the individual access via a web interface to all parameters of a product component. Building on this, it is not only possible to generate a product component file or a feasible machining plan of one single product component in the later KPP process, but rather individual modified product components as instances.

Givehchi et al. [26, 27] have introduced a method for handling an MFB containing product design and feature information that could be processing in a DPP environment. They have shown that similar product component features can be summarized and grouped as a nested directed graph of generic setups. They give a simple product component example (see Fig. 13.7) which is produced from a block of aluminum raw material. For this purpose, all the important process steps and product component features were extracted from local data sources and summarized unsorted as already mentioned above.

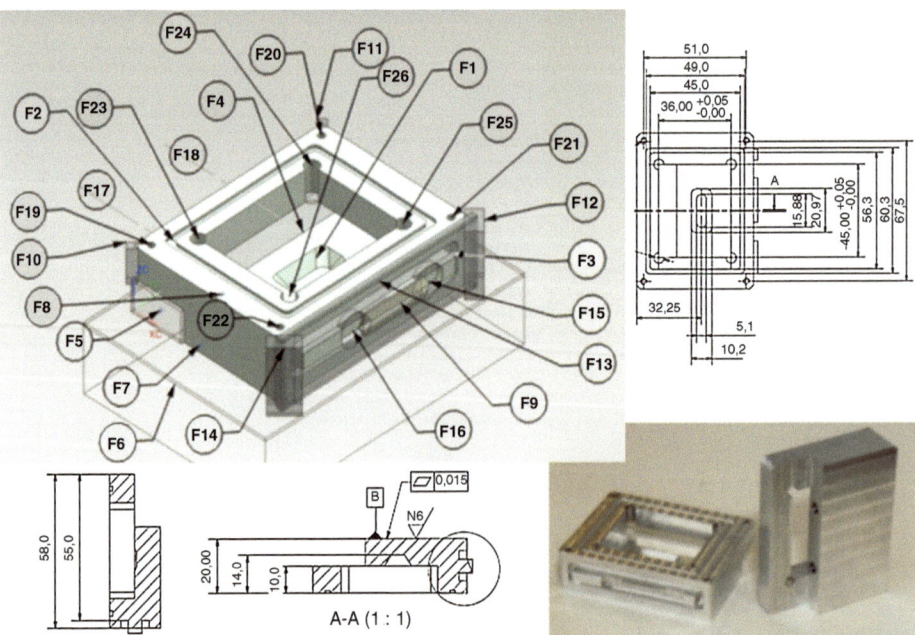

**Fig. 13.7** Exemplar product component model [26, 28]

To support the development of a KPP PoC prototype, first of all the existing implementations of WPIM were analyzed and their technologies and approaches were checked for their relevance. For the practical implementation of the KPP prototype, an implementation concept based on a web-based client/server architecture was developed. It is a modularized approach that uses the **Representational State Transfer (REST)** [29] application programming interface to integrate it into the underlying **Ecosystem Portal (EP)** which is based on **TYPO3** as its web-based front-end technology [27]. The storage and management of data are implemented independently of the data input and output. Thus, two independent software parts were created (see Fig. 13.8).

The goal of this approach was that several clients (i.e., web-based front-ends) can access the same database (back-end server) and the data management is separate from the task of data retrieval and representation. Therefore, it is possible to use any technology or programming language for the front-end, but it can still be integrated into the EP in the background. With this architecture, the front-end and the backend are no longer on the same system (see Fig. 13.8). To manage the required data, a relational database is used in the background. The front-end was developed with the help of **PHP Hypertext Preprocessor** since it is widely used and available under a free license.

At the current stage of implementation, mainly the front-end and back-end as well as the core of KPP, the process editor, is developed. The Mediator Architecture has not yet been completely integrated into the KPP application (see Fig. 13.9). Furthermore, the concept of KPP was designed for multilingualism and a user administration was conceived and implemented. However, a first functional prototype has been developed and implemented.

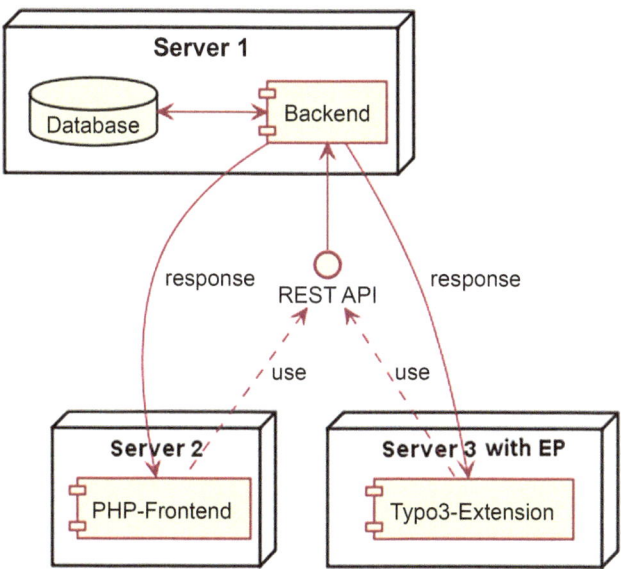

**Fig. 13.8** KPP client/server architecture with two front-ends [30] based on REST and EP technologies

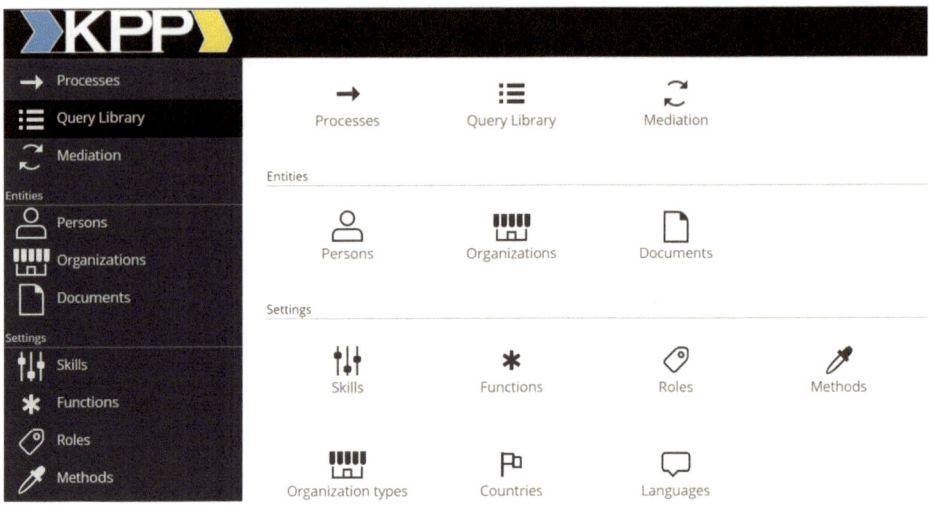

**Fig. 13.9** Screenshot of the KPP prototype's web-based user interface front-end

### 13.3.1 Ontology and Query Library

The ontology for the KPP application is derived from the WPIM ontology and adapted to KPP and the domain of production planning. It encompasses other entities such as processes, elements, people, organizations, documents, skills, functions and roles, as well as methods and tools. The ontology was conceived and created so that changes or extensions can be carried out at any time.

A query library was introduced to save and execute semantic queries. Queries are introduced here as entities, even if they only form a simple listing of semantic queries on the ontology. At this point, we encountered the fundamental problem that relational databases do not contain semantic information. This means that connections (foreign key relationships, comparisons between entities, etc.) are formulated together with the query. Furthermore, views with the connections are generated or stored procedures provide contiguous data, but there is no fundamental link between the relations (see Fig. 13.10).

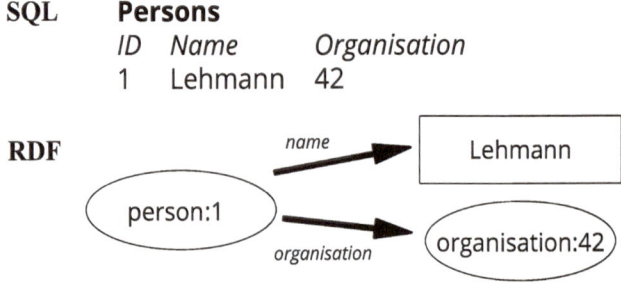

**Fig. 13.10** Schematic illustration of a translation using SQL and RDF [30]

The World Wide Web Consortium (W3C) published a paper [31] in 2004 dealing with the mapping of relational queries into semantic queries and describing a possible solution. For this purpose, another server is introduced, which runs parallel to the web and database server, that converts semantic queries into SQL queries in real time using a mapping file and returns a result.

### 13.3.2  KPP Process Editor

The centerpiece of the KPP application is a visually direct-manipulative process editor (see Fig. 13.11). This editor is based on the XML application *Business Process Model and Notation (BPMN)* [32], which is standardized for process modelling by the **Object Management Group** [33]. More specifically, we use the framework **bpmn.io** [34], which has been developed in an Open Source project since 2014. The goal of this project is to develop a framework that allows the viewing and modelling of BPMN in the web browser. It is well-documented and also extensible. However, it already meets many of the requirements, e.g., the graphical representation of process flows, annotations, and storage of graphical elements.

Furthermore, processes can be instantiated and the concept of **Lessons Learned** (already mentioned in WPIM) can be implemented by the BPMN structure because the IDs of the elements in a process instance are identical to the IDs in master processes. Thus, it is clear, which elements must be replaced. The process editor makes it possible to create complex processes, subprocesses, and process flows. It has been extended by KPP annotations, so it is possible to assign individual process steps to specific organizations, persons,

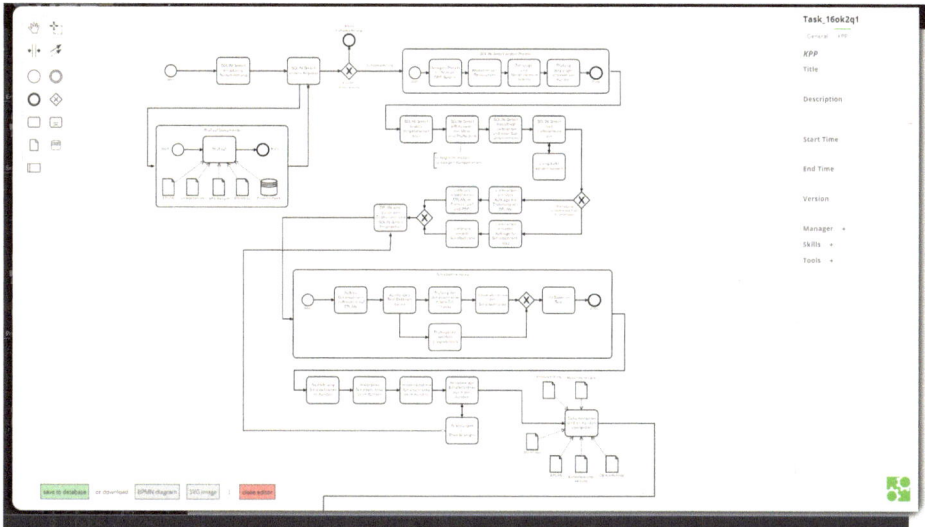

**Fig. 13.11**  Screenshot of KPP's visually direct-manipulative process editor

groups, abilities, tools, and much more. In addition, documents from a repository can be inserted into processes and can also be annotated.

### 13.3.3 KPP Mediator Implementation Approach

At this point of development, only the first Mediator, (SPPM), was initially considered. In this initial step of production planning, mainly meta and raw data are processed. That means, usually CAD-, STEP-, IGES-, and XML-data are used to derive the typical required dimensions, machines, tools, and skills for the product to be produced (Specifications for Figs. 13.6 and 13.7). For this reason, we have focused on the two most important formats in production: STEP and IGES. For the development of a suitable wrapper, we have used the full range of commands and identifiers of several examples to create a database for a global format. RDF is used as the central and global data format. In the later development, all wrappers specifically all mediators convert the various different local data into RDF. This transfer makes it possible to make semantic queries to the local data sources using SPARQL.

The sequence of such a query is designed in a way that all relevant raw files are assigned to the mediation and the user automatically receives the possible identifiers, e.g., drillings, dimensions, etc. out of these files in real time (see Fig. 13.12). With this information, he is

## Mediation

Please select a Project:

| IPhone Cover | ▾ |

Please Enter a SPARQL query:

Select ?var1 WHERE {?var1 is PRODUCT}

[ Display Identifiers ]   [ Send Query ]

These terms and relations may be used in a SPARQL statement:

| Terms (Identifiers) | Relations | Already send Queries (click to reuse) |
| --- | --- | --- |
| ADVANCED_FACE | is | Select ?var1 WHERE {?var1 is SURFACE_STYLE_USAGE} |
| APPLICATION_CONTEXT | contains | Select ?var1 WHERE {?var1 is SURFACE_STYLE_USAGE} |
| APPLICATION_PROTOCOL_DEF | count | Select ?var1 WHERE {?var1 is PRODUCT_CONTEXT} |
| AXIS2_PLACEMENT_3D | | Select ?var1 WHERE {?var1 is PRODUCT_CONTEXT} |
| B_SPLINE_CURVE_WITH_KNOTS | | Select ?var1 WHERE {?var1 is PRODUCT} |
| CARTESIAN_POINT | | Select ?var1 WHERE {?var1 is PRODUCT} |
| CIRCLE | | Select ?var1 WHERE {?var1 is PRODUCT_DEFINITION} |

**Fig. 13.12** Mediation with SPARQL-Query

able to formulate a typical SPARQL-Query and can query the desired information. For example, a user could make the request "*Select ?var1 WHERE {?var1 is PRODUCT}*", and obtains a list of the names of the individual products contained in the files. As already mentioned, the development and implementation of the mediators and wrappers are still prototypical.

## 13.4   Demonstration of KPP

In this chapter, we presented the theoretical methodology of KPP and the first prototypical PoC implementation of an exemplar application solution, its mediator, and its process editor. In the following, we will show how KPP works and demonstrate the purpose and possibilities based on a playful sandbox approach.

In our sandbox approach example, a simple miniature staircase can be produced using toy building block as components which can in turn be produced using a 3D printer. Therefore, the production planning can be carried out using the KPP PoC prototype. To do this, a new KPP project must be created and the various information and data of this project must be determined. For example, in the case of the miniature staircase, these data are the 3D models of three different types of toy building blocks as well as the blueprint of the staircase (see Fig. 13.13), the feature-based design, the machining technology and constraints, as well as the main manufacturing plan. This necessary information and raw data will be collected and recorded out of different formats and from various sources. These sources are assigned to an MFB and then integrated into the staircase project and handed over to the SPP, where the production data will be processed.

The individual sources are divided into individual tasks. The mediator functionality is used to normalize the various sources and to convert them into a uniform syntax. In this way, the data is retrievable for humans and the illusion appears of querying one single, integrated database with one integrated schema. At the same time, the individual process

**Fig. 13.13**  Simple miniature staircase with three different toy building blocks

**Same Step spacing, Design and Composite construction**

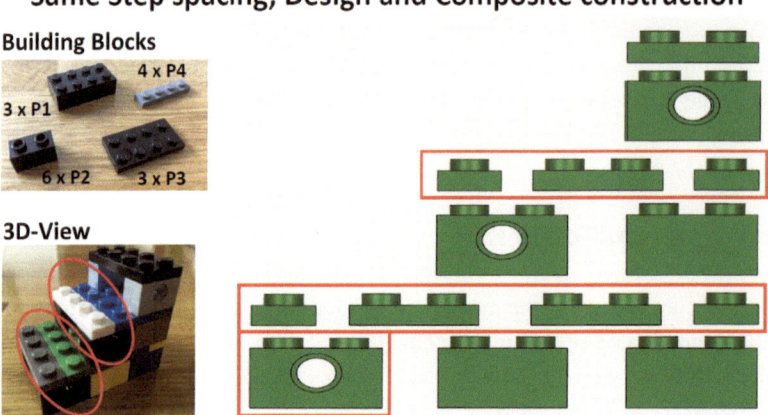

**Fig. 13.14** Simple miniature staircase with same step spacing and composite construction

can be defined in the process editor. A start and an end point must be created, and individual process and sub-process steps must be defined and linked together. The individual steps can then be annotated with explicit knowledge from the sources as well as implicit or expert knowledge from employees. Finally, all information and results are again stored in an MFB and transferred to the second process step, the ECPP.

This step runs similar to the SPP, but the information from the first step will be expanded and provided with new data. The goal is to obtain an executable machine code that can be directly executed on a suitable machine. A simple schedule will be created that 3D-prints the blocks one after the other. The data can also be annotated in this step.

In addition, the instantiation of a process will also be applied in this step. Let us assume, we would like to optimize the current staircase (e.g. equal height of steps) or enlarge it (four instead of three stair steps), then all the information of the normal staircase from SPP is still completely correct and usable as a basis for the new staircase. However, we could, for example, introduce a new part and thus adapt the height of the steps, the design and connect every individual step with each other (see Fig. 13.14).

Finally, all information and results are stored in an EFB executable on a suitable machine but not optimized on the entire shop floor (production area). This EFB will be transferred to the third process step, the OPP.

The last step, the OPP, runs similar to the other two previous steps. However, in this step, all current data from the given shop floor and basic dependencies will be introduced. For example, block 1 cannot be produced before block 2, but block 3 can be produced simultaneously with or before block 1, or another example: there are three matching 3D printers and one of them is already in use. In this way, a first optimization can be achieved because bricks can be printed in parallel on two printers, or two different blocks can be drawn simultaneously sequence by sequence on the same printer. Therefore, the output of OPP is an OFB – an optimized executable code considered all given machines and circumstances within the shop floor. Through that, all machines are appropriately utilized and not left idle.

## 13.5   Recommendations

We have introduced a knowledge-based production process planning method based on a three-level mediation. It supports semantic representations of production-knowledge integrated into a machine-readable process formalization. Furthermore, we have presented a corresponding exemplar prototypical application solution as a working example and reference implementation. We summarized our method as well as the related developments and experiences and demonstrated it by means of applying our approach to a playful sandbox example. We will now conclude with the following corresponding recommendations:

1. Production planning of a product should always be done in three phases: Concept Planning, Rough Planning, and Detailed Planning.
2. Production planning of a product has to support collaborative and knowledge based planning approaches and should be able to quickly adapt to changing circumstances.
3. Depending on the type of production and the material of the product, there are already different existing standards, machines, and formats for the various production planning sub-domains. This "raw" production planning data should be normalized into a common machine-readable format.
4. The planning of a product is basically always the organising and sequencing of specific events and processes, regardless in which production sub-domain. Therefore, an ontology for production planning should contain, on the one hand, a static and general part for the actual production planning and on the other hand a dynamic part for the special features of the production sub-domain, e.g., from the plastic- or metal-working industry.
5. We recommend checking for existing ontologies before developing a custom ontology especially for the particularities of the various domains of production planning.
6. Semantic representations of knowledge integrated into a machine-readable process formalization is a key enabling factor for sharing such knowledge in cloud-based repositories.
7. The usability and user experience of the semantic application should always be heeded to increase the user acceptance.

## 13.6   Conclusion and Future Work

This chapter has presented the current state of the art for KPP and introduced an innovative KPP method to support CAPP in the domain of manufacturing. This was demonstrated on the basis of semantic process representations, producing and consuming function blocks, and other relevant planning resources in the domain of manufacturing planning. Furthermore, we have demonstrated the prototypical implementation of KPP as well as the direct manipulative process editor and a first implementation of the mediator technology with a semantic integration including a query library based on the KPP ontology.

First, we will continue to work on the practical implementation of the methods as well as the complete integration and merge the three-level mediator and wrapper architecture together with the process editor. Thus, the complete three-level mediator architecture can be practically demonstrated. Second, based on several sample data sets, we will also demonstrate the query, search, and representation of information for the domain of manufacturing planning. Furthermore, we will evaluate the KPP method by integrating it into the overall RPP process. Therefore, future production planning activities should already be enabled to take advantage of lessons learned during setup and pre-testing as a means of optimization of production planning.

## References

1. Wiederhold G (1992) Mediators in the Architecture of Future Information Systems. The IEEE Computer Magazine, 25(3):38–49
2. Miltner F, Vogel T, Hemmje M (2014) Towards knowledge based process planning support for CAPP-4-SMEs: problem description, relevant state of the art and proposed approach ASME 2014 International Manufacturing Science and Engineering Conference (MSEC) Research Conference, Vol. 1 - Detroit, Michigan, USA, June 9–13, 2014
3. Wang L, Feng HY, Cai N (2003) Architecture design for distributed process planning. J Manuf Syst 22:99–115
4. International Organization for Standardization (2011) ISO International Standard 10303-210:2011 Industrial automation systems and integration – product data representation and exchange – part 210: application protocol: electronic assembly, interconnected and packaging design. 2011, Geneva, Switzerland
5. Vogel T (2012) Wissensbasiertes und Prozessorientiertes Innovationsmanagement WPIM – Innovationsszenarien, Anforderungen, Modell und Methode, Implementierung und Evaluierung anhand der Innovationsfähigkeit fertigender Unternehmen, Dissertation, Hagen
6. Wang L, Adamson G, Holm M, Moore P (2012) A review of function blocks for process planning and control of manufacturing equipment. J Manuf Syst 31(3):269–279
7. Wang L, Jin W, Feng HY (2006) Embedding machining features in function blocks for distributed process planning. Int J Comput Integr Manuf 19:443–452
8. International Organization for Standardization (2016) ISO International Standard 18828-2:2016 Industrial automation systems and integration – standardized procedures for production systems engineering – part 2: reference process for seamless production planning. 2016, Geneva, Switzerland
9. ProSTEP iViP Association e.V. https://www.prostep.org
10. Recommendation – Reference process for production planning PSI8, ProSTEP iVIP, March 2013. http://www.prostep.org/en/medialibrary/publications/recommendations-standards.html
11. Gernhardt B, Vogel T, Givehchi M, Wang L, Hemmje M (2015) Supporting production planning through semantic mediation of processing functionality, vol 1. International Conference on Innovative Design and Manufacturing (ICIDM), Auckland

12. International Electrotechnical Commission (2005) IEC 61499-1 Function blocks – part 1: architecture. 2005, Geneva, Switzerland
13. Lewis R (2001) Modelling control systems using IEC 61499 – applying function blocks to distributed systems. The Institution of Electrical Engineers, London. ISBN: 0852976 796
14. International Electrotechnical Commission (2003) IEC 61131-3 Programmable controllers – part 3: programming languages. 2003 Geneva, Switzerland
15. Vogel T, Hemmje M (2006) Auf dem Weg zu einem Wissens-basierten und Prozess-orientierten Innovationsmanagement (WPIM) – Innovationsszenarien, Anforderungen und Modellbildung. In: KnowTech 2006. CMP-WEKA-Verlag, Poing
16. Cyganiak R, Wood D, Lanthaler M, Klyne G, Carroll J, McBride B (2014) RDF 1.1 concepts and abstract syntax. W3C Recommendation 25 February 2014, World Wide Web Consortium (W3C). http://www.w3.org/TR/rdf11-concepts/. Last accessed 13 Nov 2014
17. SPARQL Query Language for RDF (2008) World Wide Web Consortium (W3C), 15 January 2008. Last accessed 2 Nov 2016
18. W3C OWL Working Group (2012) OWL 2 web ontology language document overview, 2nd edn. W3C Recommendation 11 December 2012, World Wide Web Consortium (W3C). http://www.w3.org/TR/owl2-overview/. Last accessed 13 Nov 2014
19. W3C (2004) OWL web ontology language overview. World Wide Web Consortium, 10 February 2004. [Online]. http://www.w3.org/TR/owl-features/. Accessed 14 Nov 2013
20. Ludäscher B, Lin K, Brodaric B, Baru C (2003) GEON: toward a cyberinfrastructure for the geosciences – a prototype for geologic map integration via domain ontologies. In: Digital mapping techniques '03 – workshop proceedings, U.S. Geological Survey open-file report 03–471
21. Bray T, Paoli J, Sperberg-McQueen CM, Maler E, Yergeau F (2008) Extensible markup language (XML) 1.0, 5th edn. W3C Recommendation 26 November 2008, World Wide Web Consortium (W3C). http://www.w3.org/TR/REC-xml/
22. Melton J (2011) ISO/IEC FDIS 9075-1 Information technology – database languages – SQL – part 1: framework (SQL/Framework), ISO Draft International Standard, ISO/IEC JTC 1/SC 32 Data management and interchange. http://www.jtc1sc32.org/doc/N2151-2200/32N2153T-text_for_ballot-FDIS_9075-1.pdf. Last accessed 13 Nov 2014
23. Robie J, Chamberlin D, Dyck M Snelson J (2014) XQuery 3.0: an XML query language. W3C Recommendation 08 April 2014, World Wide Web Consortium (W3C). http://www.w3.org/TR/xquery-30/
24. Motik B, Cuenca Grau B, Horrocks I, Wu Z, Fokoue A, Lutz C (2012) OWL 2 web ontology language profiles, 2nd edn. W3C Recommendation 11 December 2012, World Wide Web Consortium (W3C). http://www.w3.org/TR/owl2-profiles/. Last accessed 13 Nov 2014
25. Lu Y, Xu X (2015) Process and production planning in a cloud manufacturing environment. ASME 2015 International Manufacturing Science and Engineering Conference, Charlotte. MSEC2015-9382
26. Givehchi M, Schmidth B, Wang L (2013) Knowledge-based operation planning and machine control by function blocks in Web-DPP. Flexible Automation and Intelligent Manufacturing (FAIM), Porto
27. Binh Vu D (2015) Realizing an applied gaming ecosystem – extending an education portal suite towards an ecosystem portal. Master thesis, Technische Universität Darmstadt, Darmstadt
28. Givehchi M, Haghighi A, Wang L (2015) Paper: Generic machining process sequencing through a revised enriched machining feature concept. Journal of Manufacturing Systems, Vol. 37, Part 2, October 2015, Pages 564-575
29. Fielding RT (2000) Architectural Styles and the Design of Network-based Software Architectures, University of California, Irvine, CA, USA

30. Kossick J (2016) Reimplementierung, Erweiterung und exemplarische Evaluation einer verteilten und kollaborativen Unterstützung für die Produktionsplanung – Translation – Reimplementation, expansion and evaluation of a distributed and collaborative support for a production planning. Bachelor thesis, University of Hagen, Hagen
31. Prud'hommeaux E (2004) Optimal RDF access to relational databases. W3C. https://www.w3.org/2004/04/30-RDF-RDB-access/. Last accessed 29 Oct 2016
32. Visual Paradigm. Business process model and notation – diagram & tools. Hong Kong. https://www.visual-paradigm.com/features/bpmn-diagram-and-tools/. Last accessed 18 Oct 2016
33. Object Management Group (ed) (2015) OMG Unified Modeling Language (OMG UML) Version 2.5. OMG. http://www.omg.org/spec/UML/2.5/PDF/. Last accessed 27 Oct 2016
34. camunda Services GmbH (2013), BPMN-JS - a web-based toolkit for BPMN modeling. Est. 2013, Berlin, Germany. https://bpmn.io/toolkit/bpmn-js/. Last accessed 17 Mar 2018

# Automated Rights Clearance Using Semantic Web Technologies: The DALICC Framework

**14**

Tassilo Pellegrini, Victor Mireles, Simon Steyskal, Oleksandra Panasiuk, Anna Fensel, and Sabrina Kirrane

**Key Statements**
1. The creation of derivative data works, i.e., for purposes such as content creation, service delivery or process automation, is often accompanied by legal uncertainty about usage rights and compliance issues with applicable law.
2. Challenges associated with clearance issues are: (1) high transaction costs in the manual clearance of licensing terms and conditions, (2) sufficient expertise to detect compatibility conflicts between two or more licenses, and (3) negotiation and resolution of licensing conflicts between involved parties.

T. Pellegrini (✉)
University of Applied Sciences St. Pölten, St. Pölten, Austria
e-mail: tassilo.pellegrini@fhstp.ac.at

V. Mireles
Semantic Web Company GmbH, Wien, Austria
e-mail: victor.mireles-chavez@semantic-web.com

S. Steyskal
Siemens AG Österreich, Wien, Austria
e-mail: simon.steyskal@siemens.com

O. Panasiuk · A. Fensel
University of Innsbruck, Innsbruck, Austria
e-mail: oleksandra.panasiuk@sti2.at; anna.fensel@sti2.at

S. Kirrane
Vienna University of Economics and Business, Wien, Germany
e-mail: sabrina.kirrane@wu.ac.at

© Springer-Verlag GmbH Germany, part of Springer Nature 2018
T. Hoppe et al. (eds.), *Semantic Applications*,
https://doi.org/10.1007/978-3-662-55433-3_14

3. Semantic processing of license information can ease the process of rights clearance in terms of cost reduction and improvement of the decision quality. However, they are not a substitute for a human expert.
4. Reliable and trustworthy semantic systems are transparent. If the user can't reproduce or retrace a given recommendation – be it from the methodologies applied by the system or the plausibility of the output – the recommendation should be rejected.

## 14.1 Introduction

Publishing data and reusing it for commercial or non-commercial purposes has become a common practice and a cornerstone of the so called digital economy [33]. IDC & Open Evidence [18] estimated in 2013 that the European Data Economy provides 6.1 million jobs in the EU28 and could almost double by the year 2020 if high-growth is ensured. Similarly, the number of organizations producing and supplying data-related products and services could reach almost 350,000 in 2020, from 257,000 in 2014, and there could be more than 1.3 million data users by 2020.[1]

New data practices stimulated by phenomena such as open data, open innovation and crowdsourcing initiatives as well as the increasing interconnection of services, sensors and (cyberphysical) systems have nurtured an environment in which the effective handling of property rights has become key to innovation, productivity and value creation. According to the OECD, the effective management of intangible assets is the primary driver of innovation in the ICT-enabled service sector and a source of competitive advantage at the macro- and micro-level [21]. This line of argument corresponds with a study conducted by Oxford Economics which argues that "insights derived by linking previously disparate bits of data can become the sparks that ignite rapid innovation" [28]. However, according to the EU Agency for Network and Information Security, the main obstacle in the digital ecosystems of the future is the legal impact of information exchange [3]. This is especially relevant in the context of the European strategy for a data-driven economy which aims to "nurture a coherent European data ecosystem, stimulate research and innovation around data and improve the framework conditions for extracting value out of data" [5]. Accordingly rights clearance to ensure legal compatibility has become a key topic in digital ecosystems as modern IT applications increasingly retrieve, store and process data from a variety of sources [15].

Clearing and negotiating rights issues is a time-consuming, complex and error-prone task. Challenges associated with clearance issues are:

---

[1] Similar trends and figures are reported for the Spanish data broker market by Granickas [13].

(1)  High transaction costs in the manual clearance of licensing terms and conditions.
(2)  Sufficient expertise to detect compatibility conflicts between two or more licenses.
(3)  Negotiation and resolution of licensing conflicts between involved parties.

The following chapters introduce the DALICC system, a software framework that solves some of these problems by applying Semantic Web technologies to the purpose of license clearance.

DALICC stands for **Da**ta **Li**censes **C**learance **C**enter. It supports legal experts, innovation managers and application developers in the legally secure reutilization of third party data.[2] DALICC allows the attaching of licenses in a machine readable format to a specific asset and supports the clearance of rights by providing the user with information about similarity and compatibility between licenses if used in combination in a derivative work. Thus, DALICC helps to detect licensing conflicts and significantly reduces the costs of rights clearance in the creation of derivative works. Figure 14.1 gives an overview over the functional spectrum of the DALICC framework.

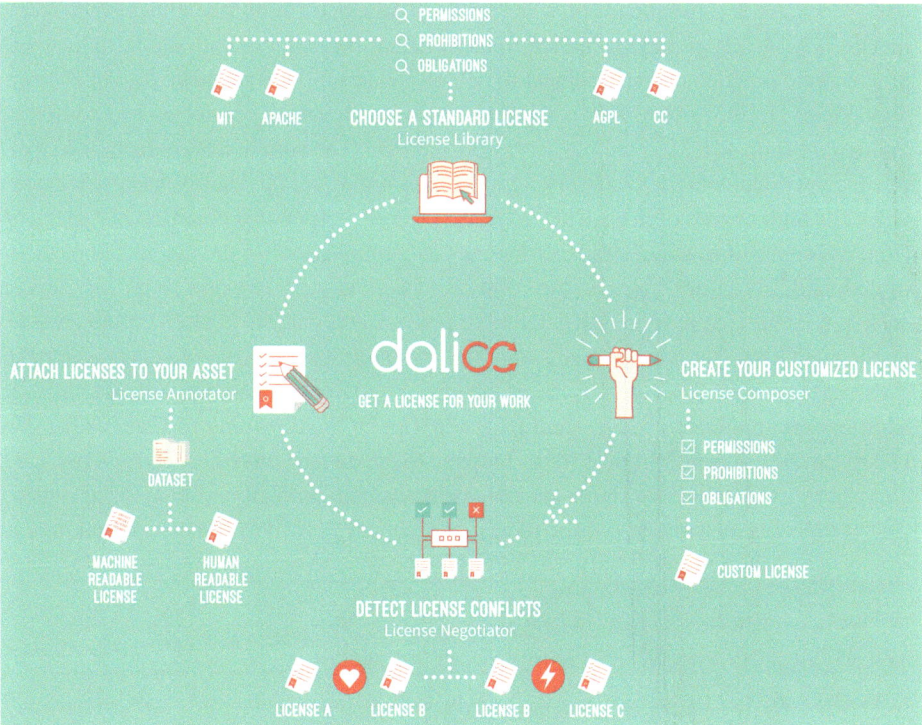

**Fig. 14.1**  The functional spectrum of the DALICC framework

---

[2]At the time of publication the DALICC service was in closed beta mode and not available to the public. A public version will be available at www.dalicc.net.

The following sections will discuss several challenges in automated license clearance and illustrate how Semantic Web technologies can be applied to solve these issues as exemplified with the DALICC framework.

## 14.2 Challenges in Automated License Clearance

### 14.2.1 License Heterogeneity

Licenses express permissions, obligations and prohibitions associated with a protectable asset as defined by copyright law or competition law. Licenses control access to, usage of, and transactions on top of digital assets, be it under conditions of property rights (all rights reserved) or public domain (no rights reserved) [34]. Figure 14.2 depicts the spectrum of available licensing models.

The growing popularity of protective and permissive licenses (some rights reserved) has added to the complexity of rights clearance in the commercial exploitation of derivative works. As a consequence, a wide array of data publishing guidelines were recommended [7, 14, 17] giving expression to the fact that licensing of data is a fairly new kind of economic practice and still subject to debate concerning the adequate design of licensing policies [1, 22, 29, 30]. This is supported by a recent survey conducted by Ermilov and Pellegrini [4] on 441,315 publicly accessible datasets. The situation is characterized by (1) insufficient documentation of licensing information (64% of all datasets had no licenses at all), (2) a high degree of license heterogeneity (more than 60 different license types), and (3) the absence of machine-readable licenses as a foundation for the automated clearance of compatibility issues.[3] Hence, the creation of derivative data works, e.g., for purposes such as content creation, service delivery or process automation, is often accompanied by legal uncertainty about usage rights and high costs in the clearance of rights issues [16]. This situation is further complicated as the efforts of license clearance increase with each additional source added to a system $[f(n) = n*(n-1)/2]$. According to Frangos [6] these efforts can be a serious obstacle for a company to create new products and services. Large companies usually operate rights clearance centres that manually evaluate legal issues in the repurposing of existing works (e.g., open source software). Such undertakings are

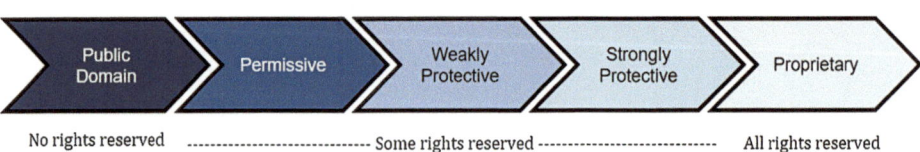

No rights reserved  ----------------------------- Some rights reserved ----------------------------- All rights reserved

**Fig. 14.2** Spectrum of licensing models

---

[3] Similar results are reported by Jain et al. [19] especially with respect to the large variety of licenses in European data clouds.

costly in terms of time and expert knowledge needed and are often out of scope, especially for small and medium sized enterprises. This is not just an obstacle to the emergence of new business models associated with data, but also slows down the rate of adoption of new data management practices, especially in the context of "a coherent European data ecosystem" as envisioned by the European Commission [5].

## 14.2.2   Rights Expression Languages

Rights Expression Languages (RELs) are a subset of Digital Rights Management technologies that are used to explicate machine-readable rights for the purposes of digital asset and access management. RELs are used to control access, explicate usage rights and govern behavioural aspects of a transaction process. Among the most prominent REL-vocabularies are MPEG-21, ODRL-2.0 (and derivatives such as RightsML), ccREL, XACML and WAC to name but a few [20]. Some RELs have a highly specific application focus, while others serve a general purpose. For example, MPEG-21 is optimized for rights management in the area of multimedia (especially digital television). On the contrary ccREL (Creative Commons Rights Expression Language) and ODRL (Open Digital Rights Language) are designed for broader application areas and have gained popularity especially in the area of content and data licensing.

Although the primary purpose of RELs is to explicate usage rights in a machine-readable form, simple tools to create, compare and process licenses attached to data assets are slowly emerging but they all have limitations.

A rudimentary version of a policy composition tool based on ODRL has been provided by the Ontology Engineering Group[4] of the University of Madrid, but this tool should be considered a proof of concept and has not been tested against real world circumstances. The same holds true for Licentia (http://licentia.inria.fr/), a license comparison tool developed by the French research institute INRIA. Also, the International Press Telecommunications Council (IPTC) working group on RightsML[5] has started to provide experimental libraries for generating RightsML and ODRL licenses in Python and JavaScript, but again, these serializations are just proof of concepts and lack a sufficient level of usability and legal validation to be suitable for commercial purposes.

Recently we have seen developments in the area of open source software that address the problem of license compatibility in the compilation of software from multiple open source libraries. For example, the Free Software Foundation provides a rich textual guide on potential licensing conflicts between open source standard licenses. This information however is not provided in a machine-readable format. Complementary to this, US-based auditing firms such as TLDR Legal (https://tldrlegal.com/) or TripleCheck

---

[4] See also http://oeg-upm.github.io/odrlapi/ & http://conditional.linkeddata.es/ldr/manageren (accessed January 4, 2016).

[5] See also http://dev.iptc.org/RightsML-Implementation-Examples (accessed January 4, 2016).

(http://triplecheck.net/index.html) have started to provide commercial services that help detecting open source licensing conflicts. What these initiatives have in common is that (1) they are specialized on software licensing, (2) compliance checking is provided as a commercial service conducted by auditing experts on top of proprietary tools, (3) none of these tools /services allows the creation of custom licenses, thus limiting the compliance check to standard licenses, and (4) no machine-readable representations of the licenses are provided to the public for advanced analytics and further reuse.

### 14.2.3 Machine-Processing of Licensing Information

Most of the work done in this research area is situated in the context of digital rights management systems and often associated with contracting issues [24, 26, 27]. Little attention has so far been paid to the issue of license compatibility and reasoning over machine-readable licensing information.

An interesting proposal for a generic logic for reasoning over licenses is provided by Pucella and Weissman [25], but it has not been implemented with existing RELs such as ODRL or MPEG-21, nor has it been evaluated in practice.

García et al. [8–10] propose an OWL ontology to describe copyright issues in closed datasets for rights clearance purposes. Their approach is based on an old version of the ODRL vocabulary and constitutes a proof of concept that has not been implemented or tested against issues arising from contemporary open data licensing.

Villata and Gandon [32] and Governatori et al. [12] describe the formalisation of a license composition tool for derivative works. They extend their research by introducing semantics based on a deontic logic [35–37] for the comparison of the permissions, prohibitions and duties stated in a given license. They also provide a demo called Licentia (http://licentia.inria.fr/) that exemplifies the practical value of such a service. This line of work is an interesting approach to detect and potentially solve licensing conflicts, e.g., by composing a new license that resolves the conflict. The pitfall of their approach is that an automatically composed license that resolves a given conflict might be logically correct but practically useless, because its conditions are either too strict or the machine-readable representation does not conform to human-readable deeds.

## 14.3   The DALICC Framework

### 14.3.1   System Requirements

According to Sect. 14.2, the following requirements can be derived for the DALICC system: (1) the output has to comply with applicable laws, (2) it needs to correctly interpret permissions, obligations and prohibitions from given licenses, (3) it must preserve abstractness and technological neutrality of the rules, and (4) it needs to support the dynamics of the rules under conditions of real world applications and usage. To achieve these goals the following problems need to be addressed:

### 14.3.1.1  Tackling REL Heterogeneity

Combining licences is simpler if all of the licences involved are expressed through the same REL. But as we have seen, various RELs have emerged for various purposes, each providing their own vocabulary and level of expressivity. Hence, it is difficult to compare licenses that have been represented by different RELs. Additionally, it can sometimes be reasonable to extend the semantic expressivity of a given REL by adding deontic expressions from other RELs to cover the requirements of a real world scenario.

DALICC solves this problem by linking vocabularies from various RELs utilising W3C-approved standards (such as RDF and OWL), thus allowing mappings between various RELs to be created. This approach allows vocabulary terms from various RELs to be combined and their expressivity to be extended beyond their original scope. Figure 14.3 illustrates a RDF graph of the standard license CC-BY utilizing vocabulary expressions from ccREL, ODRL and the DALICC vocabulary.

This RDF graph is used as input to the reasoning engine described in Sect. 14.3.1.3.

```
@prefix odrl:<http://www.w3.org/ns/odrl/2/> .
@prefix : <https://dalicc.net/license-finder> .
@prefix dalicc: <https://dalicc.poolparty.biz/DALICCVocabulary>.
@prefix rdf:<http://www.w3.org/1999/02/22-rdf-syntax-ns#> .
@prefix cc:<http://creativecommons.org/ns#> .
@prefix dct: <http://purl.org/dc/terms/> .
@prefix foaf: <http://xmlns.com/foaf/0.1/> .

:CC-BY_4.0 a odrl:Policy;
odrl:permission [
            a odrl:Permission;
            odrl:target
            <http://purl.org/dc/dcmitype/Dataset>,<http://purl.org/dc/dcmitype/Sound>,<http://purl.org/dc/dcmitype/Text>,
            <http://purl.org/dc/dcmitype/Image>,<http://purl.org/dc/dcmitype/MovingImage>;
            odrl:action odrl:distribute, odrl:reproduce,odrl:extract, odrl:derive, odrl:present;
            odrl:duty [
            a odrl:Duty;
            odrl:action cc:SourceCode, dalicc:royaltyFree, dalicc:irrevocable, dalicc:worldwide, cc:Notice,
            dalicc:noWarrantyNotice, dalicc:modificationNotice, cc:attributionName
            ]
];
odrl:prohibition [
            a odrl:Prohibition;
            odrl:target
            <http://purl.org/dc/dcmitype/Dataset>,<http://purl.org/dc/dcmitype/Sound>,<http://purl.org/dc/dcmitype/Text>,
            <http://purl.org/dc/dcmitype/Image>,<http://purl.org/dc/dcmitype/MovingImage>;
            odrl:action odrl:ensureExclusivity, dalicc:sublicense, dalicc:charge
];
dct:title "Attribution 4.0 International"@en ;
dct:alternative "CC BY 4.0";
dct:publisher "Creative Commons";
foaf:logo <http://i.creativecommons.org/l/by/4.0/88x31.png> ;
dct:source <http://creativecommons.org/licenses/by/4.0/> ;
cc:legalcode """https://creativecommons.org/licenses/by/4.0/legalcode"""@en.
```

**Fig. 14.3**  RDF-representation of CC-BY using ccREL, ODRL & DALICC vocabulary extensions

### 14.3.1.2 Tackling License Heterogeneity

Is it possible to combine a GPL Documentation License (as used by Wikipedia) with the Italian Open Government Data License v.1? Is CC-BY-ND compatible with UK-CROWN? In the creation of derivative works, the simplest approach is to only combine content under the same well-known licence. But this approach is over-restrictive, as many licences permit the licensed content to be combined. It is, however, difficult to judge whether it is permitted and how the resultant content should be licensed. There may still be subtleties arising from unclear definitions of terms (e.g., "open" or "commercial use"), special clauses (e.g., share-alike) or implicit preconditions (e.g., "everything not permitted is forbidden" or "CC0 apart from images – see restrictions in further links").

DALICC resolves these issues by producing an audited set of machine-readable licenses utilizing a given set of permissions, obligations and prohibitions – also called deontic expressions. Thus, DALICC is able to compose the crucial actions of existing standard licenses (like Creative Commons, Apache, BSD or GPL) but also allows the creation of customized licenses if none of the standard licenses suit the user's demand. To achieve the necessary level of semantic expressivity, an indepth analysis of deontic expressions in existing licenses has been conducted, matched against the vocabulary of existing RELs and complemented with additional properties, so that a sufficient level of expressivity could be reached.[6] Figure 14.4 illustrates some of the new properties DALICC utilizes to represent an arbitrary license in a machine-readable format.

By providing a sufficiently expressive set of deontic expressions, DALICC is able to represent licensing terms at a highly granular level, identify equivalent licenses and point

| Properties Affecting The Asset | Properties Affecting The License |
|---|---|
| | » dalicc:addStatement |
| » dalicc:charge | » dalicc:attributionNotice |
| » dalicc:sublicense | » dalicc:attachOffer |
| » dalicc:promote | » dalicc:chargeOffer |
| » dalicc:publish | » dalicc:irrevocable |
| | » dalicc:modificationNotice |
| | » dalicc:noWarrantyNotice |
| | » dalicc:patentFree |
| | » dalicc:patentNotice |
| | » dalicc:perpetual |
| | » dalicc:royaltyFree |
| | » dalicc:worldwide |

**Fig. 14.4** DALICC vocabulary extensions for expressing actions associated with assets or licenses

---

[6]The following 14 commonly used licenses have been semantically represented: CC-BY, CC BY-SA, CC BY-ND, CC BY-NC, CC BY-NC-SA, CC BY-NC-ND, BSD (2 clause), BSD (3 clause), MIT, APACHE, GPLv3, GPLv2, AGPL and LGPL.

the user to potential conflicts if licenses with contradicting conditions are to be combined in a derivative work. These licenses lay the foundations of the DALICC Framework, and provide the grounding for its functional components:

- The License Library lets the user select either from a set of standard licenses or customized licenses provided to the library by users.
- The License Composer employs the license models to create customized licenses.
- The License Negotiator processes and interprets the semantics encoded in the licences and checks compatibility, detects conflicts and supports conflict resolution.
- The License Annotator provides a machine-readable and human-readable version of the license that can be attached to an asset.

The mechanisms for these functional components are described in the next section.

### 14.3.1.3  Compatibility Check, Conflict Detection and Neutrality of the Rules

A common problem with semantic translation between schemas (such as RELs) is in making sure that the meaning of different terms are aligned. However, it is difficult to demonstrate the equivalence of classes, properties and instances. For RELs, the major problem arises for the instances, e.g., the precise definitions of "non-commercial", "distribution", "share-alike" etc. The classes and properties are usually simple concepts and very similar. Not all RELs support all classes though: some ignore "Jurisdiction" or even "End-user" according to the needs of the market they were developed for. To a certain degree, this will be resolved by applying Semantic Web standards, but mapping alone cannot solve the issue. More elaborated techniques, such as reasoning and inference mechanisms, are necessary to improve the accuracy of conflict detection.

To solve these issues, DALICC applies a knowledge graph that is comprised of three components: (1) a set of defined actions representing permissions, obligations and prohibitions, (2) the RDF representation of these actions, and (3) a dependency graph representing the semantic relationship between the defined actions. The core function of the knowledge graph is to encode the expert knowledge about the implicit and explicit semantic dependencies between actions. Following the work of Steyskal and Polleres [31], the corresponding dependency graph represents hierarchical relationships (e.g., use includes reproduce), implications derived from a specific action (e.g., sell implies charge) and contradictions between specific actions (e.g., non-derivative contradicts derivative). Figure 14.5 illustrates the interplay of the various components within the DALICC knowledge graph.

The DALICC reasoning engine is based on the POTASSCO suite of grounders and solvers for Answer Set Programs [11] and uses ODRL policies to detect potential conflicts in licensing terms.[7] Policies should be understood as a set of rules derived from the RDF graphs of the licenses. Herein, a rule that permits or prohibits the execution of an

---

[7]Answer Set Programming has sparked theoretical interest from the Semantic Web Community before [23] because of its relationships to the well-known SPARQL querying language.

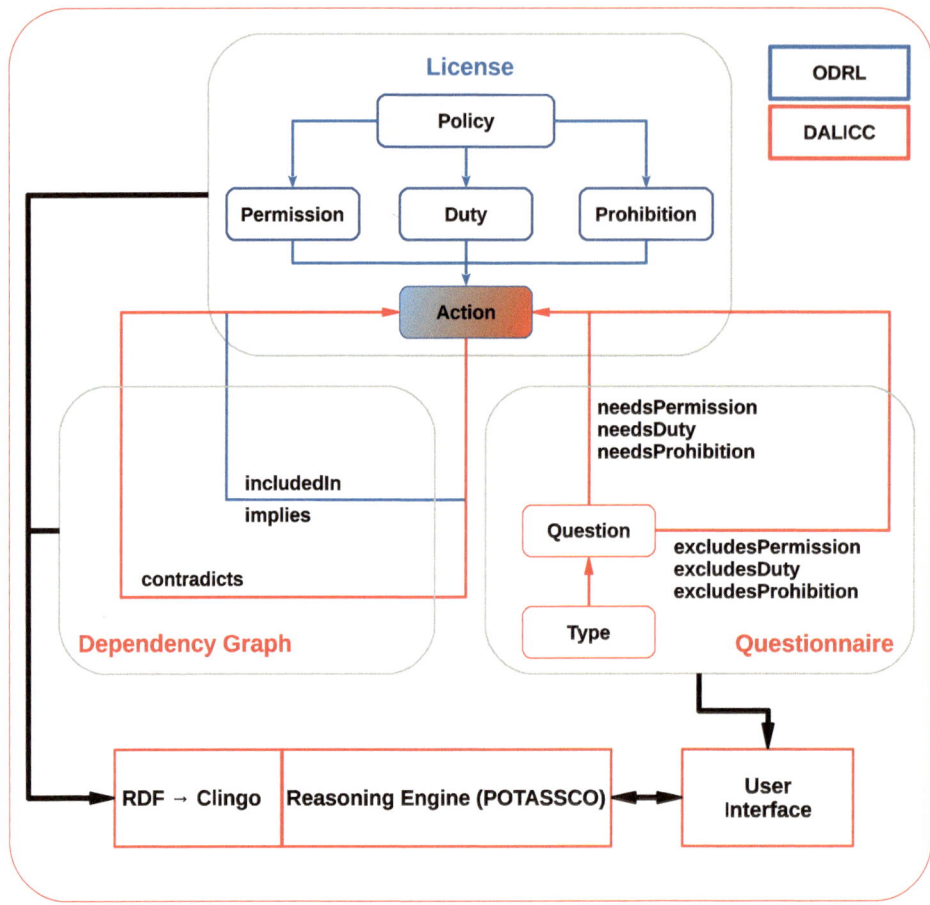

**Fig. 14.5** DALICC knowledge graph overview

action on certain assets does not only affect other rules that govern the execution of the same action on the same asset(s) but also those permitting or prohibiting related actions on the same asset(s). DALICC utilizes an RDF-to-CLINGO[8] translator to translate the given rules to a processable format and wrap it into a web service that also allows SPARQL queries. In this sense, CLINGO is not only an alternative to extensive materialization, which in this case is essential for search, but also enables listing sets of compatible statements. This latter possibility is necessary for effective computation of conflicts between licences, in particular for identifying the conflicting and non-conflicting parts of a license.

---

[8] Clingo is an ASP system to ground and solve logic programs. See also https://potassco.org/ (accessed December 12, 2017).

#### 14.3.1.4   Legal Validity of Representations and Machine Recommendations

The semantic complexity of licensing issues means that the semantics of RELs must be clearly aligned within the specific application scenario. This includes a correct interpretation of the various national legislations according to the country of origin of a jurisdiction (e.g., German Urheberrecht vs. US copyright), the resolution of problems that are derived from multilinguality (e.g., multiple connotations of "royalties" within German jurisdiction as "Lizenzgebühr", "Honorar", "Tantiemen", "Abgabe", etc.) and the consideration of existing case law in the resolution of licensing conflicts (e.g., Versata vs. Ameriprise)[9].

To tackle these issues, legal experts from inside and outside the DALICC consortium checked the legal validity of machine-readable licenses and the output of the reasoning engine for compatibility with applicable laws. In several iteration cycles the DALICC output has been tested against laws and jurisdictions, checked for its semantic accuracy and adjusted accordingly.

### 14.3.2   DALICC Implementation and Services

The DALICC user interface is based on the widespread content management system Drupal (https://www.drupal.org). Back office services are implemented with the PoolParty Semantic Suite (https://www.poolparty.biz/), which (1) manages a set of questions that guide the user in selecting a licence according to his/her needs and (2) maintains the knowledge graph which incorporates legal-expert knowledge about the licence clearing domain. The DALICC framework additionally provides a SPARQL endpoint which enables quick communication with the VIRTUOSO triple store (https://virtuoso.open-linksw.com/) containing the RDF data, the reasoning engine and the user interface. The architecture is designed to provide four basic services to the user as depicted in Fig. 14.6:

(1) **The License Library** contains machine-readable and human-readable representations of licenses. It can be accessed either via a full text search – best suited if the user already knows which license they needs – or by a faceted search that allows the user to filter licenses according to specific criteria such as asset type (e.g., data set, content or software), permissions, obligations or prohibitions. By default, the License Library is populated with the most important software and data licenses currently available. According to a recent study by Ermilov and Pellegrini [4],these include CC0, CC-BY, CC-NC-SA, UK-OGL, DL-DE-BY-1.0, IODLv2, APACHE, BSD, GPL and MIT to name but a few. Over the course of time, the library will be extended with additional licenses that frequently appear in the data domain or which are of specific importance for future applications (e.g., national open data licenses). The DALICC system also allows users to provide their customized license to the library for further reuse.

---

[9] See also Versata, Trilogy Software, Inc. and Trilogy Development Group v. Ameriprise, Ameriprise Financial Services, Inc. and American Enterprise Investment Services, Inc., Case No. D-1-GN-12-003588; 53rd Judicial District Court of Travis County, Texas.

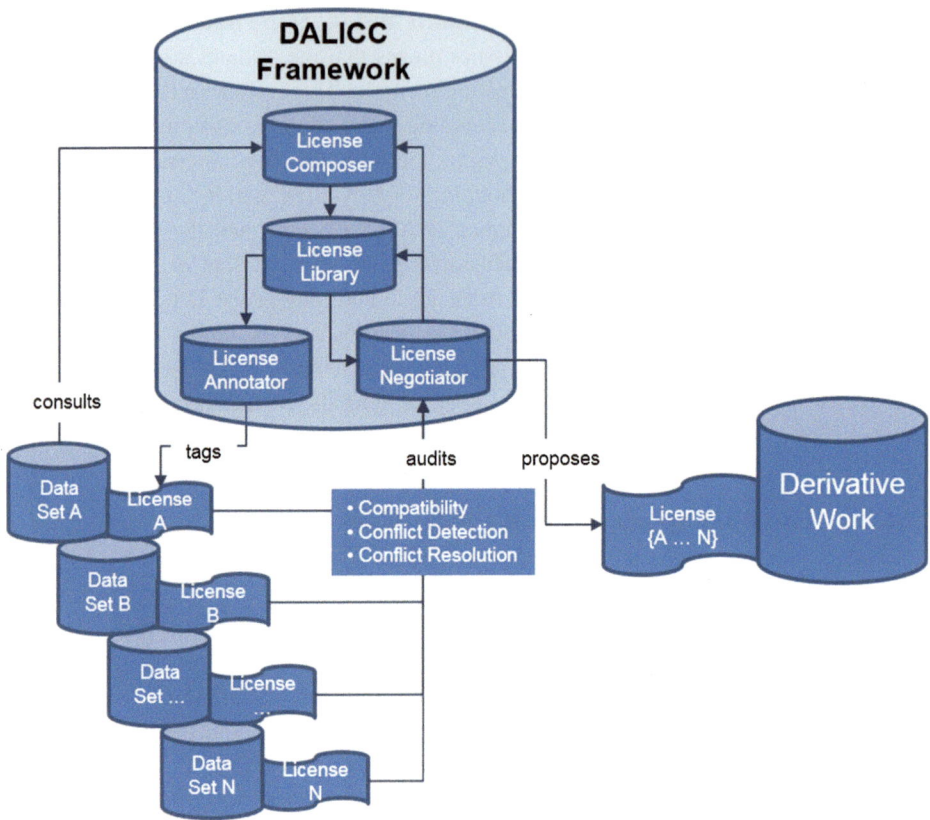

**Fig. 14.6** DALICC service architecture

(2) *The License Composer* provides the user with a simple service that allows the declaration of necessary provenance information about the asset and guides the user through a relevant set of questions that need to answered to compile a legally valid license. The composer uses ODRL, ccREL and the DALICC vocabulary as a baseline vocabulary for the specification of licensing terms. The user is additionally provided with comprehensive explanations about specific terms so that non-experts are able to understand the legal impact of their decisions and acquire the needed literacy to compile a license that suits their purposes in the wide spectrum between open and closed licensing.

(3) *The License Negotiator* is DALICC's core component. It caters for reasoning over licenses taking into account the specific context of the application provider. The negotiator checks the logical coherence of the created license, provides information on equivalence, similarity and compatibility with other licenses and supports conflict resolution between licenses. Identified resolution strategies, i.e. for re-establishing compatibility among a set of licenses, do not solely refer to choosing the most restrictive license at hand and thus potentially reducing the usefulness of the resulting (combined) license. Instead it proposes a semantically equivalent and legally sound alternative license that might resolve the detected conflict.

(4) ***The License Annotator*** finally allows exporting and/or attaching a machine-readable and human-readable license to an asset. This can be done for either standard licenses (e.g., CC-BY) already available in the License Library or for customized licenses created with the License Composer. Each newly created license can also be added to the License Library, thus allowing incremental growth of the repository and the associated knowledge base. The licenses are also be available in various formats and provided as open data to foster maximum reuse.

## 14.4 Recommendations

The DALICC framework should be understood as a supporting infrastructure for the cost-effective clearance of rights issues, thus contributing to a significant reduction of transaction costs in the commercial exploitation of derivative works. Nevertheless, it is not intended to and never should replace the knowledgeable and critical human expert on the subject matter. Users of the DALICC system or similar services utilizing semantic technologies to support critical decisions, should be aware of the following things:

- Whenever you publish a derivative work and attach a license to it, you will be held accountable. Even if you have the best intentions, make sure that the assets you built your work upon have not violated other's intellectual property. Just having a license in place does not mean that prior clearance has taken place.
- Machine-readable representations of complex intellectual constructs will never capture and resemble the semantic accuracy given in a natural language text. Hence, machine-recommendations always come with a scope open to interpretation. Recommenders should be understood as decision support mechanisms but never be taken for granted.
- Reliable and trustworthy semantic systems are transparent. If you can't reproduce or retrace a given recommendation – be it from the methodologies applied by the system or the plausibility of the output – reject it.

## 14.5 Conclusion

Licensing in general and rights clearance in particular are complex topics that require a high level of problem awareness and legal expertise. Due to the abstractness and complexity of the topic, non-legal professionals need to invest a lot of time and/or money to acquire this knowledge and search for viable solutions. Semantic Web technologies are a viable means to create systems that reduce the complexity of the subject matter and provide services that can support stakeholders at various levels of expertise to engage in and contribute to emerging digital ecosystems.

Despite the new and exciting technological opportunities semantic technologies offer to us, it is still important to stress that technology should never replace the human expert. Hence, DALICC should be understood as a supporting service in the accountable and ethical usage of property rights, to provide people with recommendations on how to protect their assets from misappropriation – be it for purposes of copyright or copyleft or something in between – and also to avoid unintentional misuse of other people's assets, that could undermine derived work.

To do so, DALICC will provide an open documentation of the system and provide its output as (linked) open data once it is fully operational. Additionally it is planned to make the DALICC framework available under a dual license, thus allowing various forms of collaborative exploitation. The framework closes the existing gap between the technological capabilities to create and publish data, and the legal infrastructure necessary to provide them on a legally secure basis for reuse. Hence, DALICC is a tool that puts data policies into practice and thus facilitates data governance. Hence, according to the data value chain provided by Deloitte [2], the DALICC framework should be understood as an enabling service for the emerging data economy.

**Acknowledgements** DALICC was funded by the Austrian Federal Ministry of Transport, Innovation and Technology (BMVIT) under the program "ICT of the Future" between November 2016 – October 2018. More information is available at https://iktderzukunft.at/en/ and https://dalicc.net/.

## References

1. Archer P, Dekkers M, Goedertier S, Loutas N (2013). Study on business models for linked open government data. ISA programme by PwC EU Services. European Commission. See also https://www.w3.org/2013/share-psi/workshop/krems/papers/LinkedOpenGovernmentDataBusinessModel. Accessed 29 Nov 2017
2. Deloitte (2012) Open growth. Stimulating demand for open data in the UK. See also http://www2.deloitte.com/content/dam/Deloitte/uk/Documents/deloitte-analytics/open-growth.pdf. Accessed 29 Nov 2017
3. ENISA (2013) Detect, SHARE, protect solutions for improving threat data exchange among CERTs, Oct 2013
4. Ermilov I, Pellegrini T (2015) Data licensing on the cloud: empirical insights and implications for linked data. ACM Press, pp 153–156. https://doi.org/10.1145/2814864.2814878
5. European Commission (2014) Towards a thriving data-driven economy. Brussels, 2.7.2014, COM(2014) 442 final
6. Frangos J (2015) New transparency in licensing: overview of the licensing facilitation act. Informed Couns 6(1):2
7. Frosterus M, Hyvönen E, Laitio J (2011) Creating and publishing semantic metadata about linked and open datasets. In: Wood D (ed) Linking government data. Springer, New York, pp 95–112
8. García R, Gil R (2009) Copyright licenses reasoning an OWL-DL ontology. In: Proceedings of the 2009 conference on law, ontologies and the semantic web: channelling the legal information flood. IOS Press, Amsterdam, pp 145–162
9. García R, Gil R, Delgado J (2004) Intellectual property rights management using a semantic web information system. In: Meersman R, Tari Z (eds) On the move to meaningful Internet systems 2004: CoopIS, DOA, and ODBASE, vol 3290. Springer, Berlin/Heidelberg, pp 689–704
10. García R, Gil R, Delgado J (2007) A web ontologies framework for digital rights management. Artif Intell Law 15(2):137–154. https://doi.org/10.1007/s10506-007-9032-6

11. Gebser M, Kaufmann B, Kaminski R, Ostrowski M, Schaub T, Schneider M (2011) Potassco: the potsdam answer set solving collection. AI Commun 24(2):107–124
12. Governatori G, Lam H-P, Rotolo A, Villata S, Auguste Atemezing G, Gandon F (2014) LIVE: a tool for checking licenses compatibility between vocabularies and data, vol 1272. See also https://hal.inria.fr/hal-01076619. Accessed 29 Nov 2017
13. Granickas K (2013) Understanding the impact of releasing and re-using open government data. In: European Public Sector Information Platform. Topic report no. 2013/08. See also http://www.epsiplatform.eu/sites/default/files/2013-08-Open_Data_Impact.pdf. Accessed 29 Nov 2017
14. Guibault LM (2011) Open content licensing: from theory to practice. Amsterdam University Press, Amsterdam
15. Hoffmann A, Schulz T, Zirfas J, Hoffmann H, Roßnagel A, Leimeister JM (2015) Legal compatibility as a characteristic of sociotechnical systems. Bus Inf Syst Eng 57(2):103–113. https://doi.org/10.1007/s12599-015-0373-5
16. Houghton J (2011) The costs and benefits of data provision. Report to the Australian National Data Service. Centre for Strategic Economic Studies, Victoria University
17. Hyland B, Wood D (2011) The joy of data – a cookbook for publishing linked government data on the web. In: Wood D (ed) Linking government data. Springer, New York, pp 3–26
18. IDC & Open Evidence (2013) European Data Market. SMART 2013/0063. See also https://drive.google.com/a/open-evidence.com/file/d/0B5Co3wBffnzhUTBQUklCS0VoRTg/view?pref=2&pli=1. Accessed 29 Nov 2017
19. Jain P, Hitzler P, Janowicz K, Venkatramani C (2013) There's no money in linked data. http://knoesis.wright.edu/pascal/pub/nomoneylod.pdf. Accessed 29 Nov 2017
20. Kirrane S, Mileo A, Decker S (2015) Access control and the resource description framework: a survey. Semant Web J. See also http://www.semantic-web-journal.net/content/access-control-and-resource-description-framework-survey. Accessed 29 Nov 2017
21. OECD (2008) Intellectual assets and value creation. See also http://www.oecd.org/sti/inno/40637101.pdf. Accessed 29 Nov 2017
22. Pellegrini T (2014) Linked Data Licensing – Datenlizenzierung unter netzökonomischen Bedingungen. In: Schweighöfer E et al (eds) Transparenz. 17. Int. Rechtsinformatik Symposium IRIS 2014. OCG Verlag, Wien
23. Polleres A, Wallner JP (2013) On the relation between SPARQL1. 1 and answer set programming. J Appl Non-Class Log 23(1–2):159–212
24. Prenafeta J (2010) Protecting copyright through semantic technology. Publ Res Q 26(4):249–254
25. Pucella R, Weissman V (2002) A logic for reasoning about digital rights. In: Proceedings of the 15th IEEE workshop on Computer Security Foundations. IEEE Computer Society, Washington, DC, pp 282–294
26. Rodriguez E, Delgado J, Boch L, Rodriguez-Doncel V (2015) Media contract formalization using a standardized contract expression language. IEEE Multimed 22(2):64–74. https://doi.org/10.1109/MMUL.2014.22
27. Rodriguez-Doncel V, Delgado J (2009) A media value chain ontology for MPEG-21. IEEE Multimed 16(4):44–51. https://doi.org/10.1109/MMUL.2009.78
28. Roehring P, Pring B (2013) The value of signal and the cost of noise. Oxford Economics, London
29. Sande MS, Portier M, Mannens E, Van de Walle R (2012) Challenges for open data usage: open derivatives and licensing. https://www.w3.org/2012/06/pmod/pmod2012_submission_4.pdf. Accessed 12 Feb 2016
30. Sonntag M (2006) Rechtsschutz für Ontologien. In: e-Staat und e-Wirtschaft aus rechtlicher Sicht. Richard Boorberg Verlag, Stuttgart

31. Steyskal S, Polleres A (2015) Towards formal semantics for ODRL policies. In: Rule technologies: foundations, tools, and applications – 9th international symposium, RuleML 2015, Berlin, Germany, August 2–5, 2015, Proceedings, pp 360–375. https://doi.org/10.1007/978-3-319-21542-6_23

32. Villata S, Gandon F (2012) Licenses compatibility and composition in the web of data. In: Third international workshop on consuming linked data (COLD 2012), Boston, Nov 2012, https://km.aifb.kit.edu/ws/cold2012/. hal-01171125

33. World Bank (2014) Open data for economic growth. See also http://www.worldbank.org/content/dam/Worldbank/document/Open-Data-for-Economic-Growth.pdf. Accessed 29 Nov 2017

34. Ball A (2014) How to license research data. A Digital Curation Centre and JISC Legal 'working level' guide. http://www.dcc.ac.uk/resources/how-guides/license-research-data. Accessed 29 Nov 2017.

35. Rotolo A, Villata S, Gandon F (2013) A deontic logic semantics for licenses composition in the web of data. In: Proceedings of the fourteenth international conference on artificial intelligence and law, ICAIL '13. ACM, New York, pp 111–120. https://doi.org/10.1145/2514601.2514614

36. Guido G, Ho-Pun L, Antonino R, Serena V, Fabien G (2013) Heuristics for licenses composition. Frontiers in artificial intelligence and applications. pp 77–86. https://doi.org/10.3233/978-1-61499-359-9-77

37. Cabrio E, Palmero Aprosio A, Villata S (2014) These are your rights. In: Presutti V, d'Amato C, Gandon F, d'Aquin M, Staab S, Tordai A (eds) The semantic web: trends and challenges. Springer International Publishing, Cham, pp 255–269

# Managing Cultural Assets: Challenges for Implementing Typical Cultural Heritage Archive's Usage Scenarios

**15**

Kerstin Diwisch, Felix Engel, Jason Watkins, and Matthias Hemmje

> **Key Statements**
> 1. Exchanging data and integrating similar collections are tasks which often occur in the Cultural Heritage domain and presents one of the biggest challenges in this domain.
> 2. A wide number of archives and collections exist in this domain. To make their contents widely accessible, linking these data inventories is a key task. This leads to the heavy usage of Linked Data and Semantic Web technologies.
> 3. The emphasis on linking and connecting data inventories demands the application of de-jure and de-facto standards in this domain.

## 15.1  Introduction

In the domain of cultural heritage, curators and archivists are often concerned with exchanging data between archives and integrating similar collections with regards to contents. Typical usage scenarios in this area can only be realized by semantically integrating all available data sources. Thus, the main difficulty is the different composition and scope of the data inventories. These might have complex structures and, in addition, are often stored

K. Diwisch (✉)
Intelligent Views GmbH, Darmstadt, Germany
e-mail: kdiwisch@i-views.com

F. Engel · J. Watkins · M. Hemmje
University of Hagen, Hagen, Germany
e-mail: felix.engel@fernuni-hagen.de; matthias.hemmje@fernuni-hagen.de

© Springer-Verlag GmbH Germany, part of Springer Nature 2018
T. Hoppe et al. (eds.), *Semantic Applications*,
https://doi.org/10.1007/978-3-662-55433-3_15

in distributed and heterogeneous data resources. Furthermore, global networking leads to multilingual contents which makes integrating the data sources even more difficult. Therefore, the main task of semantic integration consists of bridging the different levels of heterogeneity of the data model, schema, semantics, language, granularity, depth, domain, and range of the data and corresponding meta-data resources by means of establishing appropriate mappings allowing a semantic integration.

Semantic Web technologies might be a solution for these problems and challenges. Ontology Matching, for example, is already successfully applied for bridging some of the heterogeneity types. However, it is not suitable for matching and therefore resolving and mapping, i.e., integrating heterogeneity conflicts at all levels. This is due to the fact that automatic techniques have not yet led to satisfactory results. In addition, most matching, mapping and resulting integration techniques are only suitable for highly formal ontologies while compositions of lightweight ontologies and schemas are actually used in practice. Thus, the matching results' quality is insufficient for the currently deployed ontologies. However, matching is crucial for the semantic integration success of the distributed data sources. As a result, a main part in semantic integration is still done manually by domain experts. It would be preferable to support their work by at least semi-automatic techniques.

Today a huge amount of standards and taxonomies are commonly used in the domain of cultural heritage. Their automatic matching and corresponding resolving of heterogeneities between them by appropriate mappings could facilitate semantic integration. Yet, each of the models and schemata proposed by the standards consists of mixing and matching of, e.g., available vocabularies, which makes recognizing interlinkings between vocabularies and the correct interpretation of their semantics the main challenges in this semantic integration.

The book chapter describes typical usage scenarios in the domain of cultural heritage archives and discusses the use of Semantic Web technologies to implement these scenarios. In addition, commonly used standards and vocabularies in this area are presented and ways to integrate these vocabularies are discussed.

## 15.2    Characteristics of the Cultural Heritage Domain

Cultural heritage means conservation and preservation of contemporary artifacts of a society. However, definitions about what might be such artifacts differ in the aspect of whether immaterial artefacts such as dances or plays are also part of a society's cultural heritage. This leads to the determination between tangible and intangible heritage [27]. Nowadays, most definitions include intangible heritage artifacts [27].

In this domain, exchanging data between archives and semantically integrating collections within a larger context provide the most workload for archivists and curators. Typical usage scenarios in this area, which will be described in the following paragraphs, can often only be realized by integrating all available data sources. The main difficulties in achieving this are the differences in composition and scope of the data inventories. They consist of complex structures and are often stored in distributed resources. Additionally, global networking leads to multilingual contents which makes the semantic integration of the data sources even more difficult. Therefore, the main tasks in constructing a cultural heritage archive and connecting it with others of the same domain consists of bridging the different levels and types of heterogeneity in the data model, schema, semantics, language, granularity, depth, domain, range of the data [10, pp. 37–39] and corresponding meta-data resources by means of establishing appropriate mappings, allowing for semantic integration [10, p. 11], [28].

There are already a wide range of collections and archives in existence in this domain. Linking these data inventories has been an active research topic in this field for several years. Projects such as Europeana [13] try to integrate heterogeneous data sources and provide single-point access to their contents by forming virtual archives or libraries. The ongoing research work in this field has also resulted in a multiplicity of data models, meta-data schemas, and corresponding vocabularies. Re-using these models, schemas and vocabularies might ease linking the different data inventories and provide ways to query a wide range of data sets at once. This strongly reminds of the whole idea of Linked Data to publish data on the web in such a way that a single global data space will be created [2]. Thus, it is no wonder that Linked Data and Semantic Web technologies are widely used in this domain.

## 15.3   Standards for Archives in this Domain

The main challenges for archives in the domain of Cultural Heritage are concerned with semantic integration of distributed and heterogeneous data sources and in particular bridging the different levels and types of heterogeneity between data inventories. Furthermore, there is the need for making the data easily accessible and enabling interoperability. This results in extensive requirements for an archive's technological architecture.

### 15.3.1 Open Archival Information System (OAIS)

ISO 14721 Open Archival Information System (OAIS) is a standard for a conceptual reference model for technical architectures of archives with similar requirements. The reference model was developed by NASA [28]. The goal of the OAIS is to "establish a system for archiving information, both digitized and physical, with an organizational scheme composed of people who accept the responsibility to preserve information and make it available to a designated community" [14]. The reference model offers approaches from simple architectures with single accessing and ingestion points, up to architectures for federated archives. Covered tasks are ingestion, packaging, certification, preservation planning, and data management. It is a purely conceptual standard and therefore does not recommend specific technologies.

However, it does define an information model based on information packages which hold information on the contained data objects, their preservation and relevant metadata. There are three types: The Submission Information Package (SIP) which is sent to the archive, the Archival Information Package (AIP) which is stored within the archive, and the Dissemination Information Package (DIP) which is sent from the archive.

The OAIS reference model also describes Functional Entities within an archive's structure and their main tasks. Furthermore, it provides descriptions for interfaces and data flows within the archive and also external connectors to data consumers and producers. Sharing functional entities through federated or cooperative archives is another topic covered in the standard.

Designing an archive's architecture around the OAIS reference model helps to consider all aspects of data collection, assembly, integration, curation, preservation and accessibility and might also reduce the effort when linking and combining archives, a task which often occurs in this domain.

### 15.3.2 Metadata Standards

In the domain of cultural heritage, there already exists a huge amount of standards and taxonomies which are commonly used. Their automatic integration could facilitate semantic integration on the data schema and data instance level. Yet, each of the schemata proposed by the standards consists of mixing and matching of the available vocabularies, which makes recognising links between vocabularies and the correct interpretation of their semantics one of the main challenges in semantic integration. Still, using these vocabularies and schemas is advisable as it eases linking and exchanging between data resources.

The existing vocabularies and schemas cover different topics and tasks. They range from base schemas up to certain meta-data areas. Table 15.1 shows a list of widely used vocabularies and schemas in this domain. It is by far neither complete nor does it even

**Table 15.1** Overview of commonly used metadata vocabularies in the Cultural Heritage domain

| Name | Type | Description |
|---|---|---|
| BIBFRAME [17] | Schema | Mostly used in libraries |
| CDWA Lite [3] | Schema | Used in the Anglo-Saxon area, contained by the Getty society |
| CIDOC-CRM [4] | Schema | Mostly used in archives and museums |
| EDM [9] | Schema | The Europeana Data Model |
| FRBR [1] | Schema | Preferentially used in libraries, but also in archives and museums |
| Dublin Core [8] | Metadata vocabulary | Annotations and provenance |
| PBCore [18] | Metadata vocabulary | Extension of Dublin Core with focus on media assets |
| FOAF (Brickley, Miller [11]) | Metadata vocabulary | Person data |
| schema.org [21] | Metadata vocabulary | Annotations for web pages |
| SKOS and SKOS-XL (Miles, Bechhofer [22]) | Schema and metadata vocabulary | Thesaurus and taxonomies, as well as annotations and provenance |

resemble the most popular vocabularies and schemas. It is intended to show the wide range of the existing vocabularies and their different layouts.

The usage of de-jure and even de-facto standards is advisable as it eases data integration issues. However, the extent of reuse depends on the usage scenarios in this domain.

## 15.4   Semantic Technologies

Typical usage scenarios in this area can only be realized by integrating all available data sources. Thus, the main difficulty is the different composition and scope of the data inventories. They consist of complex structures and are often stored in distributed resources. Besides, global networking leads to multilingual contents which makes integrating the data sources even more difficult. Therefore, the main task consists of bridging the different levels of heterogeneity as language, semantics, depth and range.

Semantic Web technologies might be a solution for these problems. Ontology Matching, for example, is already successfully applied for bridging some of the heterogeneity levels and types. However, it is not suitable for all levels and types. This is due to the fact that automatic matching techniques have not yet led to satisfying results. In addition, most matching techniques are only suitable for highly formal ontologies while compositions of lightweight ontologies and schemas are actually used in practice. Thus, the matching results' quality is not sufficient for the currently deployed ontologies. However, this is crucial for the semantic integration's success. As a result, a main part in data integration is still done manually by domain experts. It would be preferable to replace these processes by at least semi-automatic techniques.

## 15.5   Typical Usage Scenarios

From expert interviews with data providers, curators and archivists of the Cultural Heritage domain we concluded the following list of important usage scenarios. Each usage scenario is described and it is discussed which standards and technologies could be applied in order to support such a scenario.

### 15.5.1   Sharing Media Files

Curators and archivists in the domain of Cultural Heritage often need to manage media assets such as videos and photographs. Often, digital archives originate from collected paper material which then is digitized by scanning documents and annotating the files manually.

Thus, sharing of media files is important in this particular domain because of its characteristics. Archiving performances or exhibitions can often be equated with archiving videos and photographs of these artefacts as these are the only source of evidence for them.

Furthermore, there might not even exist textual content for an artefact, or related resources might contain little information (e.g., posters or announcements for an exhibition). This leads to the special importance of annotations. Even though video and image analyzing algorithms have improved over recent years, finding a certain scene in a play or detecting a special characteristic in a painting heavily relies on annotations and metadata.

Annotating data and creating metadata can be supported with the help of Semantic Web technologies. Certain de-facto standard vocabularies such as Dublin Core [8], FOAF (Brickley, Miller [11]), SKOS (Miles, Bechhofer [22]) and MPEG 7 [15] provide schemas for these needs. By using them, not only the creation of metadata can be enhanced, also the linking between archives is easier if they use the same standard vocabularies.

However, annotations are not the only challenge. Sharing media files also requires good performance and a high bandwidth of the archive's interfaces in order to serve consumers' requests in an acceptable timeframe. Semantic Web interfaces are known for lacking performance. Therefore, caching Linked Data contents and providing other interfaces with semantic integration, feature extraction and indexing technologies (e.g., Apache Solr, Apache kafka and Apache Flink [24]) [6] are often required to provide adequate performance. Thus, a semantically integrated media content interface needs to be carefully designed around Semantic Web as well as Data Analysis and Data Integration technologies.

### 15.5.2 Archiving Content Data in Textual Form

Even though sharing media files is of high importance for an archive in the Cultural Heritage domain, providing content in textual form also plays an important role. Interview records, stage directions, and painting interpretations contain important textual information for archive consumers such as scholars. In order to make this data easily accessible when searching an archive, text analyzing algorithms and part-of-speech tagging are often used to extract metadata from content data.

In this scenario, ontologies are often used [6]: they can serve as the base for classification, tagging, text extraction, Named Entity Recognition and Named Entity Linking [23]. However, often gazetteers and lists are adequate for these tasks and require less initial effort. The choice depends on the user stories and use cases whether the effort of using an ontology is worthwhile (see Chap. 2). For example, if the archive's data will be used in a Linked Data context, using LD vocabularies for annotations and metadata might be useful for easier sharing and reducing effort in preparing data sets.

### 15.5.3 Providing User-Friendly Data Sharing Interfaces

Providing consistent interfaces based on common standards reduces time and effort for data producers, especially for those who provide new data on a regular basis (e.g., uploading photographs after each performance), as they merely need to assimilate their upload

**Fig. 15.1**  The Submission Information Package (SIP) is sent from the data producer to the archive

facilities to certain standards and might even be able to reuse their interfaces because of shared standards. Thus, archivists and curators who rely on external data producers make an effort to provide user-friendly data sharing interfaces not only to reduce time and effort on the producer's side but also to reduce their own efforts in importing and reworking data sets.

In this scenario, it is also advisable to check the description of advised data flows in the OAIS reference model regarding interfaces for data producers on the application layer [25] and define the Submission Information Package (SIP) which is sent from the data producer to the archive as shown in Fig. 15.1.

Sources for collecting and ingesting data are not limited to data producers as already defined. Importing data from Social Networking Systems and Data Sharing Platforms has become a rather common usage scenario [20]. Thus, an easy and manageable way to record this data is needed.

A common technique for automatically sharing and ingesting data is the use of application profiles which define the rules for valid content and data structure [19]. In the most basic form, guidelines for manual data entries with data ingestion definitions (e.g., a list of mandatory attributes and their accepted data formats) could serve as an application profile. However, the use of machine-readable application profiles reduces the manual importing effort by providing ways to automate validation of new content. Application profiles in the RDF format also support the linking between existing and new content, for example, through reasoning. The following listing shows a part of a Description Set Profile, a constraint language for Dublin Core Application Profiles, in RDF [16].

```
:dTemplate_video rdf:type :DescriptionTemplate;
  dsp:minOccur "0"^^xsd:nonNegativeInteger;
  dsp:maxOccur "Infinity"^^xsd:string;
  dsp:resourceClass :Video;
  dsp:statementTemplate :litST_title,
    :litST_mediaPosition,
    :nonlitST_embodimentPerformance,
    :nonlitST_depicts.
```

The constraints contain information on the number of individuals of the class ":Video" and describe which predicates can be used for these entities. Within the importing process, data sets on videos could then be parsed and validated according to these constraints.

### 15.5.4 Make Archive Contents Discoverable

**A huge effort in the media archiver's regular workload is dedicated to** making data findable and accessible. The time spent on this work heavily relies on the quality of the metadata and annotation data. Another factor is the range and variety for user's search options: Is their search restricted to names and dates, or can they scan through the whole content? Can they find a video of a certain performance by looking for "red shoes", or do they have to provide the artist's name and creation date in order to find a certain painting?

There exists a wide variety of metadata vocabularies in RDF. Some of them are already extensively used in this field and are described in the prior paragraph. The proximate possibility to link contents among different data sources and automatically derive data through reasoning makes the use of Semantic Web technologies advisable in this usage scenario. This is further supported by the potency of SPARQL, the query language for RDF, for querying and manipulating RDF data stores [12, p. 202].

Furthermore, a number of ontologies for describing provenance have evolved over the last decade (e.g., Dublin Core Dublin Core Metadata Initiative [8], or Open Provenance Model [26]) [7]. Provenance data plays an important role in successful querying and searching data. Thus, the use of these ontologies could enhance the search quality. However, annotation and metadata vocabularies also exist in XML. Even some of the provenance vocabularies exist in XML (e.g., Open Provenance Model). Therefore, the use of RDF is not always necessary, and which technology is most apt must be decided by the archive's provider.

Semantic Web technologies can also be used on the users' access side: Semantic annotations and ontologies can be used to help understanding and processing textual search requests. Again, this is a performance issue. Thus, search quality and response time have to be balanced.

### 15.5.5 Interlinking Between Archives

An OAIS [14], in general, is a self-contained and independent entity. However, there exist several reasons to consider interaction between OAISs and their applied technical or organisational interlinkage, e.g., in order to make their archive's data accessible to a wider designated community, archivists and curators can link their archive with other archives of the same domain. For example, a collection which consists of paintings from Spanish artists could be linked to an archive with data from an exhibition on Pablo Picasso. Users of both applications could benefit from having access to increased amounts of data, allowing them to provide more data to their end users without having to extend their collection or archive's content.

Another benefit of linking collections and archives is the possibility to compare data sets. For example, provenance of data could be validated through comparing it with a similar data set from a different collection. Inconsistencies could also be found with this method.

In its current version, the OAIS ISO foresees four categories of associations between OAISs to share content and functionality: *Independent, Cooperation, Federated and Shared Resources.*

In *Shared Resources*, two or more OAISs will share a set of OAIS *Functional Entities* (FE, see Sect. 15.3.1), e.g. to reduce costs related to their realisation and maintenance. However, even in this case, the operation of all functions that are required to form a full OAIS *FE* remains an extensive task.

Regarding this, a change request has been issued within the review. If it will be approved, the OAIS ISO will introduce so-called *Virtual Archives* (VA) with *Distributed Functional Entities* (DFE). Therein, a VA is an agreement between OAISs to link or integrate their distributed functionalities with each other in a complementary way. This will essentially enable the integration of different functionalities, from different OAISs to build the full functional body of a OAIS FE.

However, sharing contents between archives is explicitly described in the OAIS ISO under the category of "Cooperating" OAISs. In this interaction category, an OAIS OAIS1 could, for example, act as a Consumer of a second OAIS, OAIS2. In a more complex scenario, OAIS1 and OAIS2 could act as Data Producers and Consumers towards a third OAIS, OAIS3. Thereby, Consumers have to separately search in all archives to find data objects of interest [25].

Essentially, sharing a Common Catalogue makes several archives accessible via a single interface. This approach is described in the OAIS interaction category "Federated". Here, the OAISs have a Common Catalogue combining the archives' metadata and annotations. Search queries are verified with the contents of the *Common Catalogue* and then processed in the corresponding archives. *Dissemination Information Packages* (DIP) matching the request will then be sent to the consumer as shown in Fig. 15.2.

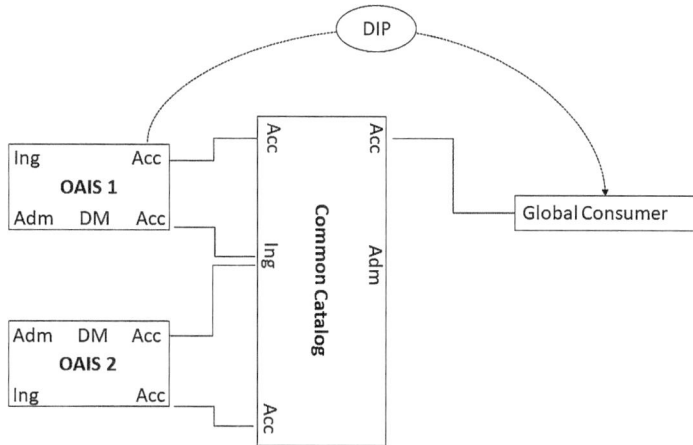

**Fig. 15.2** After accessing the Common Catalog, a Dissemination Information Package (DIP) with the requested data contents are sent to the consumer by one of the affiliated OAIS

There are however several issues with the federation of OAIS as well. Amongst others, unique identifiers have to be applied for non-conflicting access, and user management has to be addressed properly amongst participating OAISs to allow only permitted access.

The OAIS reference model has not yet adapted the concept of federated archives for submission methods. However, federated archives could certainly use a common ingestion interface and receive mutual *Submission Information Packages* (SIP) for updating their archival storage.

One way to apply Semantic Web technologies is to generate the Common Catalogue through the means of Ontology Matching. Ontology Matching is the process of finding similar concepts or entities within different ontologies, rate their correspondence, and store the matching results in an alignment [10]. The following listing shows an excerpt from an Ontology Matching result using the Alignment API [5]:

```
<Cell>
<entity1 rdf:resource="http://ontology1.owl#Person"/>
<entity2 rdf:resource="http://ontology2.owl#Artist"/>
<measure
rdf:datatype="http://www.w3.org/2001/XMLSchema#float">0.75</measure>
      <relation>=</relation>
</Cell>
```

The listing shows a match between the entities Person and Artist which is rated with a 0.75 correspondence.

Such an alignment could be used as the base for a Common Catalogue. This would provide links between the heterogeneous schemas and would therefore reduce the manual workload for creating the Common Catalogue.

## 15.6   Recommendations

What can be done to deal with the challenges of semantically linking archives in the Cultural heritage domain and bridging all levels of heterogeneity?

1. Utilize standards. As there already exists a wide range of other archives and collections in this domain, using standards helps to reduce the effort of integrating and linking your archive with other sources.
2. Check which metadata vocabularies are prevalent in your particular realm and make use of them.
3. While the usage of Semantic Web technologies is encouraged, as it is absolutely suitable for this domain, it should be used with caution as, for example, performance issues might occur.

## 15.7  Conclusion

The Cultural Heritage domain provides a lot of challenges for curators and archivists. The main challenges are data integration and making data widely accessible. As there already exist a high number of archives and collections, linking and exchanging data between heterogeneous data inventories is another task archivists and curators are concerned with.

These challenges can be overcome by using de-jure and de-facto standards in this domain. Designing an archive in compliance with the OAIS reference model, for example, is helpful for later integration of other archives or data sources. Comprising an archive's schema of existing vocabularies and applying commonly used vocabularies for annotating data supports exchanging data and providing standard interfaces.

The domain's environment supports the usage of Linked Data and Semantic Web technologies. However, they do not always provide the best solution. The usage scenarios should dictate the technologies to be used.

## References

1. Bekiari C, Dörr M, Le Boeuf P (2008) FRBR. Object-oriented definition and mapping to FRBRER. International Working Group on FRBR and CIDOC CRM Harmonisation. Available online at http://www.ifla.org/files/assets/cataloguing/frbrrg/frbr-oo-v9.1_pr.pdf. Checked on 26 Jan 2016
2. Bizer C, Heath T, Berners-Lee T (2009) Linked data-the story so far. In: Semantic services, interoperability and web applications: emerging concepts. pp 205–227. IGI Global, Hershey, Pennsylvania
3. CDWA Lite (2016) Available online at http://www.getty.edu/research/publications/electronic_publications/cdwa/cdwalite.html. Updated on 1 Feb 2016, checked on 15 May 2017
4. CIDOC CRM (2017) Available online at http://www.cidoc-crm.org/. Updated on 15 May 2017, checked on 15 May 2017
5. David J, Euzenat J, Scharffe F, Trojahn dos Santos C (2011) The alignment API 4.0. Semant Web J 2:3–10
6. Deuschel T, Heuss T, Humm B, Fröhlich T (2014) Finding without searching: a serendipity-based approach for digital cultural heritage. In: Proceedings of the digital intelligence (DI) conference, Nantes, 17–19 Sept 2014
7. Ding L, Bao J, Michaelis JR, Zhao J, McGuinness DL (2010) Reflections on provenance ontology encodings. In: McGuinness DL, Michaelis JR, Moreau L (eds) Provenance and annotation of data and processes. Third international provenance and annotation workshop, IPAW 2010. Troy, NY, USA. Springer, Berlin/Heidelberg, pp 198–205
8. Dublin Core Metadata Initiative (2012) DCMI metadata terms. Available online at http://dublincore.org/documents/dcmi-terms/. Checked on 29 June 2017
9. Europeana Data Model Documentation (2014) Available online at http://pro.europeana.eu/page/edm-documentation. Checked on 15 May 2017
10. Euzenat J, Shvaiko P (2013) Ontology matching, 2nd edn. Springer, Berlin
11. FOAF Vocabulary Specification (2014) Available online at http://xmlns.com/foaf/spec/. Updated on 14 Jan 2014, checked on 15 May 2017

12. Hitzler P (2008) Semantic Web. Grundlagen. Springer, Berlin/Heidelberg (eXamen.press). Available online at https://doi.org/10.1007/978-3-540-33994-6
13. Isaac A, Clayphan R, Haslhofer B (2012) Europeana: moving to linked open data. Inf Stand Q 24(2/3):34–40
14. ISO 14721:2012. Space data and information transfer systems – Open Archival Information System (OAIS) – Reference model. Available online at https://www.iso.org/standard/57284.html. Checked on 15 May 2017
15. Kosch H, Heuer J (2005) MPEG-7. Gesellschaft für Informatik (GI). Available online at https://gi.de/service/informatiklexikon/detail/mpeg-7/. Updated on 4 Nov 2017, checked on 11 Nov 2017
16. Nilsson M (2008) Description set profiles: a constraint language for Dublin Core Application Profiles. Available online at http://dublincore.org/documents/dc-dsp/. Checked on 15 May 2017
17. Overview of the BIBFRAME 2.0 Model (2010) BIBFRAME – Bibliographic Framework Initiative, Library of Congress. Available online at https://www.loc.gov/bibframe/docs/bibframe2-model.html. Checked on 15 May 2017
18. pbcore.org – Public broadcasting metadata dictionary project. Available online at http://pbcore.org/. Checked on 15 May 2017
19. Reinking K (2013) Einsatz eines Dublin Core Application Profile im digitalen Archiv der Pina-Bausch-Stiftung. Bachelorarbeit. Hochschule Darmstadt, Darmstadt
20. Salman M, Buechner MFW, Vu B, Brocks H, Becker J, Heutelbeck D, Hemmje M (2016) Integrating scientific publication into an applied gaming ecosystem. GSTF J Comput (JoC) 5(1). Available online at http://dl6.globalstf.org/index.php/joc/article/view/1650. Checked on 19 Sept 2016
21. schema.org. Available online at http://schema.org/. Checked on 15 May 2017
22. SKOS Simple Knowledge Organization System Reference (2009) Available online at https://www.w3.org/TR/2009/REC-skos-reference-20090818/. Updated on 18 Aug 2009, checked on 15 May 2017
23. Swoboda T, Hemmje M, Dascalu M, Trausan-Matu S (2016) Combining taxonomies using Word2vec. In: Sablatnig R, Hassan T (eds) Proceedings of the 2016 ACM symposium on document engineering, DocEng 2016, Vienna, 13–16 Sept 2016, ACM, pp 131–134. Available online at http://doi.acm.org/10.1145/2960811.2967151
24. The Apache Software Foundation (2017) Apache projects. Available online at http://www.apache.org/index.html#projects-list. Updated on 27 Apr 2017, checked on 15 May 2017
25. The Consultative Committee for Space Data Systems (2012) Reference model for an Open Archival Information System. Available online at https://public.ccsds.org/pubs/650x0m2.pdf
26. The Open Provenance Model (2011) Available online at http://openprovenance.org/. Updated on 27 June 2011, checked on 15 May 2017
27. Vecco M (2010) A definition of cultural heritage. From the tangible to the intangible. J Cult Herit 11(3):321–324
28. Visser U, Schuster G (2002) Finding and integration of information – a practical solution for the semantic web. In: Web semantics: science, services and agents on the World Wide Web. pp 74–79. Elsevier, Amsterdam.

# The Semantic Process Filter Bubble

**16**

## Christian Fillies, Frauke Weichhardt, and Henrik Strauß

**Key Statements**
1. Models of business processes should be automatically annotated while modelling.
2. Knowledge-based navigation using semantic tagging creates a filtering bubble, supporting model readers in accessing relevant information.
3. Process context information allows for efficient recommendations of the available knowledge space elements.
4. Tagging information of model elements creates context information that can be used for document recommendation based on semantic search mechanisms.

## 16.1 Introduction

As information search in corporate environments becomes more cumbersome with every new storage location and every new application that creates data for a specific purpose, we introduce the concept of knowledge-based navigation as an alternative way to find relevant information. The objective is to gather information from business processes to allow better navigation through larger collections of models and pertaining information. The concept also targets identification of the types of data eligible for navigation purposes as well as the creation of a user-friendly application to display the information.

The concept has been implemented in intranet environments, which will also be described in the course of this chapter.

C. Fillies (✉) · F. Weichhardt · H. Strauß
Semtation GmbH, Potsdam, Germany
e-mail: cfillies@semtalk.com; hstrauss@semtalk.com; fweichhardt@semtalk.com

© Springer-Verlag GmbH Germany, part of Springer Nature 2018
T. Hoppe et al. (eds.), *Semantic Applications*,
https://doi.org/10.1007/978-3-662-55433-3_16

### 16.1.1 Process Concepts in the Information Society

The word "process" has different meanings. In this chapter, we look at it in a knowledge management context: We would like to address business processes. Business processes can be described in several ways: Textually, using tables, via a picture or using a model. Models can contain pictures, but also describe the process in a logical way.

In Business Process Modelling, currently there are two main use cases: On the one hand, process models are created for technical reasons, on the other hand, they are created for documentary or organisational purposes. With respect to technical process modelling, the use case is "graphical programming", where an application (a so called workflow) is created from the model. This kind of process model needs a lot of detailed technical information in order to be able to create a running application (see [1, 2]. It is not suitable as a documentary or organisational view of the process, that would allow a non-technical reader to understand how the process has to be performed. It is also not possible to get a general idea of interconnections between several processes looking at those mentioned technical models (workflows).

Documentary or organisational process modelling is used to prepare for certification audits, to support organisational design, to define requirements for new systems, in the context of quality management and also for knowledge management purposes. These models are typically published within the organisation in some kind of portal application.

The concept of knowledge-based navigation is based on the description of business processes, also called process models. Process models in this sense not only describe events and steps of a process, but also the involved systems, roles, documents and other data.

While modelling, elements of process models are connected to other content available in the relevant knowledge space in order to allow users to access this content via a portal in a precise way. One way of connecting to other content is tagging via the assignment of business objects. Business objects are the core elements on which a business process is executed. During process modeling, business objects will automatically be assigned to process steps and consequently to the business process model. Roles are also associated. Using all this context information, a filtering bubble is created, helping users to access information in a purposeful way.

In the following text, we focus on documentary models (business processes), which are usually used in Knowledge Management contexts.

### 16.1.2 Current Issues

Business process models are often used to describe organizational collaboration [3]. For this reason, they include the assignment of additional information such as guidelines, forms, responsibilities, risks etc., to their corresponding process elements. This information is often part of other collections of information as well. The entirety of this information (process models and assigned information) forms a complex structure of linked data and graphical elements. Supportive tools are necessary to purposely navigate through these structures.

Usually, process models are published on a process portal site, e.g. on a Microsoft SharePoint Server, or by generating an HTML representation. These representations can become large and complex, meaning that it can become difficult to find specific information. Therefore, tools to find or filter information are necessary. Some search and navigation support may already be in place, such as filtering based on user roles or workflow functionality, to retrieve relevant information for specific tasks that are executed. Searching for information in a more general way is possible as well. Up until now, what has been missing is a solution that combines these different approaches so that they benefit from each other. Workflows and business processes are two different things, as described above, so normal search functionality is not able to find information contained in both representations. An approach that combines these aspects will maximize user satisfaction and minimize the time needed to find relevant information.

End-user tasks are the main focus as they make up the foundation for the information needed for work processes which are then decomposed into their specific work steps.

### 16.1.3  Organisational and Technical Requirements

In order to be able to develop a process navigation system, there are some prerequisites that we found necessary in our projects:

- All employees in an organization are assigned to one or more specific roles in at least one active business process.
- Roles are assigned to specific process tasks during the creation of a business process model.
- Process models are enriched semantically by using modelling tool features which contain associated documents, hyperlinks, document classes and other information, e.g. business objects.
- The identification of a specific user in the business process is given. He or she is either determined manually or given by a workflow application. This context will then be used for contextual information filtering.
- Annotation of process elements with tags in the form of business objects will act as the foundation for semantic enrichment. As a consequence, it is possible to deduce the relevant business objects for each role. This is also true for information like documents or hyperlinks.

### 16.1.4  Technical Platform

The solution has been built using the business process modelling product "SemTalk®", developed and distributed by Semtation GmbH. It is an object-oriented process modelling tool with an integrated tagging mechanism. It is designed to create business process models, organizational charts or ontology models and allows for tagging of process models while modelling (based on a given or user created vocabulary). It is distributed as a Microsoft Visio graphics application plug-in based on .NET Framework technology so that SemTalk works synergistically with other products based on Microsoft technology.

SemTalk can be used to create purpose-oriented representations of different types of processes. Ultimately, SemTalk is a tool used to better understand business processes so that they can be changed or optimized. Optimization is achieved through in-depth knowledge of all processes and information essential in everyday company life. A SemTalk process model is more than just a graphical representation, it is a collection of information, roles, IT systems etc., and also includes links to relevant documents. Data that is to be included in models is the choice of the process creator and this data is highly configurable and flexible. New elements can be added without much effort. Created information and process elements in SemTalk become objects that can be reused multiple times, as well as in different process models. In that way, the modeling process reflects the distinct world of a company's language.

In order to fulfil the above-mentioned requirements, SemTalk enables, among others, the following capabilities:

- The ability to conclude semantic data from process models in a way that improves navigation.
- Automatic assignment of business objects to process elements.
- Integration of semantic search and parametrization.
- Identification of context data that describe the current end-user's work context.
- Creation of a user-friendly portal site to display the information.
- Creation of simple-to-use functions to make it easy to understand and to quickly find relevant information.

In our scenario, Microsoft SharePoint Server is utilized as the publishing platform for the process models. Usually, a process portal site is created, which contains a graphical viewer for all processes along with some SharePoint functionality blocks (web parts) for displaying the information attached to each process step. Process models are saved on a SharePoint Server. In addition, associated model information that is created while modeling processes can be saved, e.g. roles, IT systems and documents. Also, prior to beginning a modeling project, it is possible to save a list of predefined terms that can be used when modeling process steps [4].

The Microsoft SharePoint workflow engine is used to automatically execute processes on SharePoint sites [5]. In our scenario, SharePoint itself is used to implement and execute processes that establish a more technical representation of real-world processes. Another important aspect of Microsoft SharePoint Server is the search engine, which is intended to find documents and other information located on a SharePoint site. The cloud-based version of SharePoint makes use of graph analysis and AI in order to recommend the users relevant information. The standard SharePoint search experience is agnostic of explicitly specified business processes and their relation to the users. Exploiting the process models allows filtering of the information in a way that it stays in the relevant "Bubble" of the process.

## 16.2   Comprehensive Description of the Portal

As already discussed, the concept of process-based navigation is based on the description of business processes. The goal of the portal is to connect documents to process elements not only using static hyperlinks but also by the usage of search results within the knowledge base (SharePoint in our case) which are parameterized based on the process context.

The Microsoft SharePoint search engine provides several features relevant to this goal, such as query configuration capabilities, taxonomies and storage for documents and structured information. SharePoint Server is a leading platform for publishing process models along with the required metadata.

Microsoft SharePoint includes a workflow engine to implement processes as executable workflow representations. The specification is done in SharePoint with a tool named SharePoint Designer. Also, a few third-party applications are available which can simplify the specification of SharePoint workflows. However, it is not just the executable process that can be interpreted as workflow, but also the unstructured user interaction in SharePoint associated with a specific business process that can be identified (e.g. the search for a document and use of forms can be regarded as a workflow, but they are just not as explicitly defined in a formal workflow). These unstructured processes usually have room for improvement. Therefore, it is important to identify those processes and their respective process steps to figure out which documents or information are needed frequently and those which are not as accessible as they should be.

### 16.2.1   Parameterisation of Search Engines Based on a Process Context

There are several options to use information for filtering and parameterisation of search engines, including [6, 7]:

- Query rules: Search terms impact the way the engine is processing the query, i.e., there can be specific actions for specific keywords or users:
  - There are sources that contain the searchable content. Query rules can dynamically decide which source to use, e.g. different SharePoint sites.
  - User segmentation: By use of user roles or other properties the query can be altered by using query rules [8]. Different users can get different search results.
- Content Enrichment: While crawling, an enrichment service connects external concepts to already identified ones.
- Custom Entity Extraction: Based on defined vocabularies, keywords are recognised in searched documents.

Applying this to our process context case, three options were identified for parameterizing search.

The first option is to use dictionaries as a source for the search within a process context (Custom Entity Extraction) [9]. In conjunction with SemTalk, these dictionaries are exported from process and data models to SharePoint. Dictionaries are composed from roles, process steps, synonyms, business objects and other model elements. While SharePoint's search engine is crawling, a search for dictionary terms in crawled documents having matching annotations is included. Annotated documents are useful to filter the relevant documents in search results.

The second option is to use the concept of Content Enrichment. Content enrichment has the same purpose as Custom Entity Extraction, but it is not based on internal dictionaries. Instead, it relies on a web service that provides annotations.

In our case, the third option is the use of process models as a basis to execute search queries:

- Which information is relevant for a specific user's role?
- In which process or process step is the user currently involved or within which process does the user try to find information?
- Which documents are attached or which document is most likely searched that is available for this specific process or process step?
- Which information is associated with the current process domain?
- Which documents have been accessed by other users having the same role or using the same workflow?

For the recommendation engine in this case, the focus was on the third alternative, as the model is the main asset here, while workflow, logging and usage data are coming from third-party tools.

## 16.2.2 Identification of Process Steps Based on Current User Activities

The problem is: How to deduce the process specific parameters in order to customize a search?

One option is to identify process steps from SharePoint workflows (workflow monitoring): We have to establish a mapping from a business process model to a technical workflow specification. In our solution, we have used the respective SemTalk functionality. This implies that workflow tasks are associated with process steps in the model in order to establish a common context. Using SharePoint's current state of a workflow instance, we can find the assigned process step in the process model. Relevant information is extracted from the process model. Business objects are often available to define a subset of information for filtering.

Another option is the analysis of SharePoint log files in order to infer actual process steps. Process mining of event data is becoming a popular practice for this task these days, but it is always a process specific project to identify and implement measuring infrastructure. Message hubs – especially in the cloud – enable us now to build more sophisticated event-to-process model mappings. Native SharePoint logging mechanisms do not render enough relevant data though, so it is necessary to use an additional logging application.

Based on a SemTalk specific event logging functionality for the process portal site, it is possible to log user interactions and their roles while they access certain processes and documents in the workflow. This log data is then used to recommend information to the next user who has a similar role and workflow task.

### 16.2.3  Process Information in a Process Portal

A process portal is made of multiple web parts. Figure 16.1 shows a basic process portal with its functionality blocks (web parts) aligned on the left and the right side of a Visio Web Access Web Part, that presents Visio/SemTalk process models within SharePoint.

The web parts on the left side of the figure allow the user to navigate within the process models. One of the web parts offers a drill down in the process hierarchy, while the other web part provides a navigation tree including all process models grouped by type and by name. This combination of graphics and structured information gives users access to those process models which are relevant to them. Filtering is done based on the roles in the process models. Web parts on the right side of the figure provide additional information regarding the specific process steps. If the user selects a shape on the Visio web access web part, the most important properties that have been assigned during process modeling such as roles, attributes, documents etc, are displayed dynamically. There are plenty of other web parts available to present information to the user. Often, Wiki pages, reports, pictures and many of the out-of-the-box web parts provided by SharePoint are used in combination with the model viewer web part.

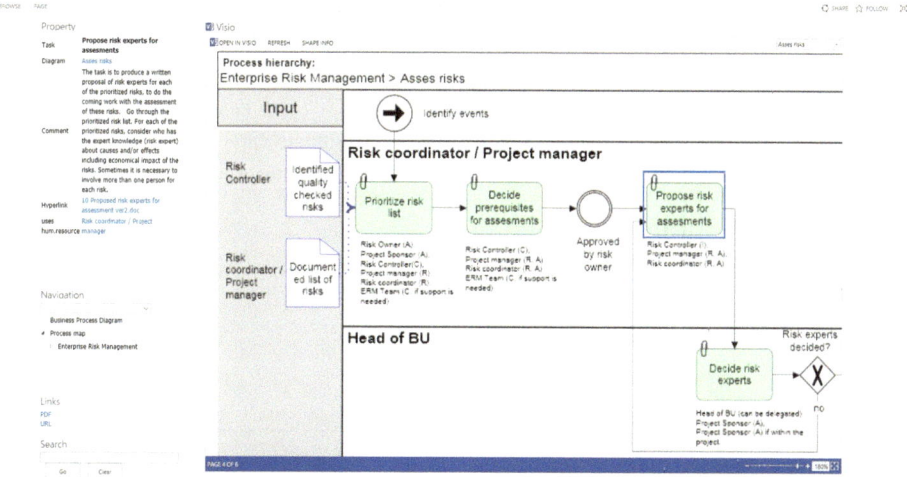

**Fig. 16.1**  Example of a process portal

## 16.2.4 Process-Context Based Portal Event Logging

Virtually all click-events will be logged to an external database that adds information about it. These are process models, activities, documents and searches together with a timestamp, and workflow tasks. This information can then be used to improve the suggestions given to future users with a similar role, who use similar search terms in the same process model. Portal interaction event logging at the process portal has several benefits:

- Information is presented according to previous interaction.
- Most significant documents are rated higher.
- Documents often searched in this specific process underline their importance for the process or process step. Based on this information the document may be added directly to the process model or will be presented with a higher ranking to the next user.
- The process portal learns from its users about relevant information.

It is important to judge the event's significance and relevance of its documents with respect to:

- Who is the current user, applying user segmentation [10].
- The workflow tasks: Which workflows are performed by the user.
- What process model is the user currently investigating .
- The role assigned to the user.
- Search terms.
- Documents opened

In contrast to the user-driven navigation in the portal, log event-based information is pre-emptive. The system automatically adds suggestions to help users find what they need.

A set of ranked documents and process models is then sent to the process portal to be displayed as suggestions.

The example given in Fig. 16.2 shows the current user on the left and a suggestion box based on previously logged events, on the right.

SharePoint online, in contrast to SharePoint 2013 – which is the foundation of this article – makes use of cloud-based web services that offer even more filtering and search functionality,

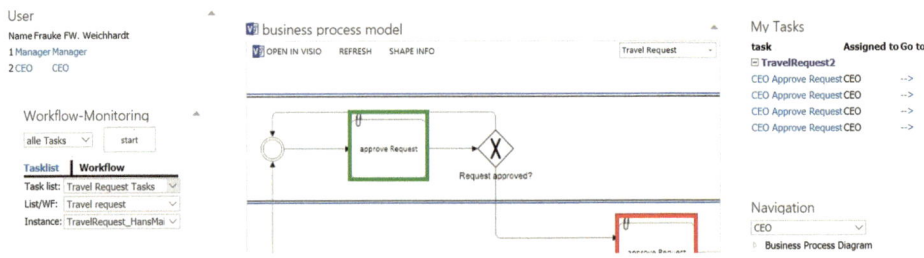

**Fig. 16.2** Example for user identification in the portal

including knowledge graphs and AI. These instruments would augment search result quality if applied to documents and model content. An ideal recommendation algorithm would combine all of them.

## 16.2.5  Workflow Monitoring

As the process portal is set up, a workflow monitoring component may be added. Workflows are connected to process models using the workflow definition file (an XML structure) that is imported into SemTalk. SemTalk extracts workflow tasks and creates usable SemTalk objects accordingly. The process modeller manually attaches those workflow tasks to applicable process activities. After the assignment is done, the mapping information between workflow task and process activity will be exported into a SharePoint list. This list is later on accessed by the process portal in order to provide the workflow monitoring view.

Using SharePoint workflows to give specific information according to current workflow task has the following benefits:

- Documents are linked directly to workflow tasks and thus to workflow monitoring, which is integrated with the process portal site.
- Information is presented next to the graphical process model.
- The workflow monitoring solution connects business process models with live workflow instances.
- Selection of a process step also selects the corresponding workflow task with its attached documents and structured data.
- The process portal is setup in order to show the business process for a specific user along with his current workflow tasks using user segmentation.
- No active search is needed.

Figure 16.3 displays a workflow monitoring portal example.

In this scenario, a user of the portal selects a specific workflow instance and the Visio Web Access web part displays the workflow as an overlay to the process model according to the previously created connection between workflow task and process step. A step with a finished workflow task appears with a green frame while activities with workflow tasks in progress get a red frame so that users can see the state of a specific workflow instance.

On top of the left side of the portal is a web part which shows the identified user. Identification of users allows for personalized filtering of workflow task and allows the recommendation of appropriate activities.

On the top right of the portal is a web part located which shows all workflow tasks for this specific user. Selection of a task in this web part will navigate to the corresponding process model and will highlight the activity to which the workflow task is attached. The user gets an overview of how his task is positioned in the context of the general business process. If he selects a shape in the Visio Web Access Web part, the user gets all attached

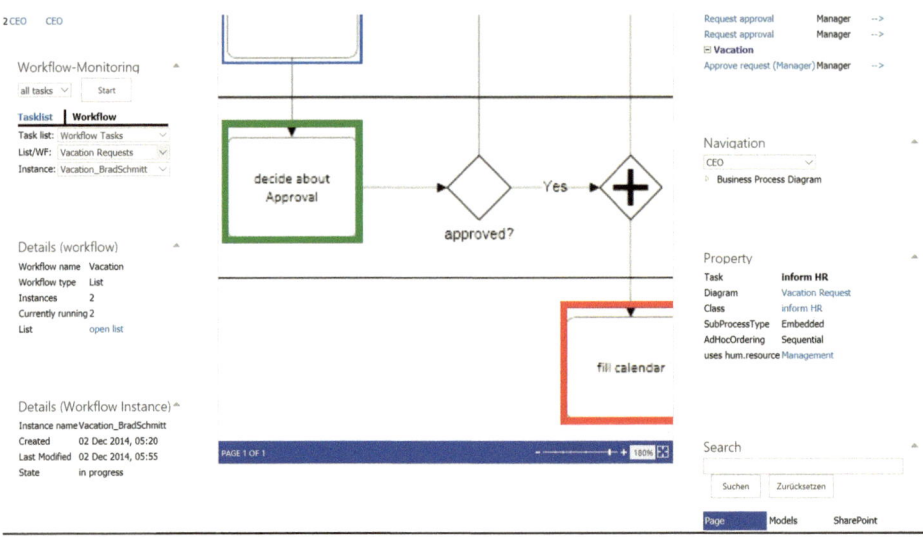

**Fig. 16.3** Example of a workflow monitoring portal

information from the process model, e.g. linked documents, roles, attributes etc. Much of the information that the user needs is presented to him preemptively without an explicit search action. He has direct access to the documents and items related to this workflow instance.

Similar portals combining the static process models with dynamic runtime information have been built e.g. task lists, project management systems, ERP systems, event processing or IoT scenarios. The overall concept is to capture the state of execution and use the visual presentation and the logical process graph to navigate to relevant content.

## 16.3 Recommendations

1. Semantic annotations may contribute to knowledge-based navigation within process models.
2. Semantic search can be used to automate document search for individual process steps. Search parameters can be based on process context information.
3. Knowledge-based navigation should be integrated in a knowledge ecosystem such as, Microsoft SharePoint. However, its concept of Managed Metadata cannot utilize the whole potential of semantic technologies.
4. Even without having the whole semantic technology stack available it is possible to build useful navigation and search solutions.
5. Semantic projects in a corporate context are always in need of domain-specific conceptual models. If you already have some, the probability that your project will be successful is much higher.

## 16.4   Conclusion and Future Research

Process-based navigation has proven to support information filtering. Adding semantic concepts reduces the information overflow and helps to focus on business-relevant information. The overall goal should be to present process relevant information to the user without having to manually trigger a search, but instead have it presented preemptively.

Future research might consist of adding other knowledge sources to the process context, so that this information also can be accessed in a more flexible way.

## References

1. Freund J (2017) BPM guide. BPMS – die nächste Generation. http://www.bpm-guide.de/2010/01/16/bpms-die-nachste-generation/
2. Wolverton M, Martin D, Harrison I, Thomere J (2008) A process catalog for workflow generation. In: Sheth A et al (eds) ISWC, LNCS, vol 5318. Springer, Berlin /Heidelberg, pp 833–846
3. Weßel C (2011) Einführung von Prozessmodellen als Chance zur Organisationsentwicklung: Das Beispiel Klinische Behandlungspfade. Manuskript. Frankfurt am Main
4. Semtation G (2014) Microsoft Visio und Microsoft SharePoint. Von http://www.semtation.de/: http://www.semtation.de/index.php/de/visiosharepoint
5. msdn.microsoft.com (2012) Get started with workflows in SharePoint 2013. Von http://msdn.microsoft.com/: http://msdn.microsoft.com/en-us/library/office/jj163917.aspx
6. Battiston F (2013) How to customize SharePoint 2013 search results using query rules and result sources. Von http://blogs.technet.com/: https://blogs.technet.microsoft.com/mspfe/2013/02/01/how-to-customize-sharepoint-2013-search-results-using-query-rules-and-result-sources/
7. technet.microsoft.com (2013) Overview of the search schema in SharePoint Server 2013. Von http://technet.microsoft.com: http://technet.microsoft.com/en-us/library/jj219669.aspx
8. Skinner R (2013) User context sensitive searching in SharePoint 2013 Part 1. Von http://richardstk.com/: http://richardstk.com/2013/07/12/user-context-sensitive-searching-in-sharepoint-2013-part-1/
9. diZerega R (2013) Advanced content enrichment in SharePoint 2013 search. Von http://blogs.msdn.com/: http://blogs.msdn.com/b/richard_dizeregas_blog/archive/2013/06/19/advanced-content-enrichment-in-sharepoint-2013-search.aspx
10. Peschka S (2012) Using user context (AKA segmentation) in search with SharePoint 2013. Von http://blogs.technet.com/: http://blogs.technet.com/b/speschka/archive/2012/12/02/using-user-context-aka-as-segmentation-in-search-with-sharepoint-2013.aspx

# Domain-Specific Semantic Search Applications: Example SoftwareFinder

# 17

Bernhard G. Humm and Hesam Ossanloo

**Key Statements**

1. Domain-specific semantic search applications may improve user experience.
2. Semantic search applications extend traditional full-text search by semantic application logic. Examples are semantic faceted search, semantic auto-suggest, and similar product recommendations.
3. The core of a semantic search application is an ontology. All semantic application logic depends on the ontology.
4. The implementation of semantic application logic is specific to the application domain and the ontology used.
5. For many specific application domains, e.g., software component search, pre-defined ontologies do not exist.
6. Where pre-defined ontologies do not exist and manual ontology development is not feasible, simple ontologies may be developed semi-automatically based on data mining.

B. G. Humm (✉)
Hochschule Darmstadt, Darmstadt, Germany
e-mail: bernhard.humm@h-da.de

H. Ossanloo
Object ECM GmbH, Braunschweig, Germany

© Springer-Verlag GmbH Germany, part of Springer Nature 2018
T. Hoppe et al. (eds.), *Semantic Applications*,
https://doi.org/10.1007/978-3-662-55433-3_17

## 17.1   Introduction

Software development today means, to a large extent, integrating existing software components: A component is a unit of software that has a published interface such as database management systems, middleware components, and GUI libraries which can be used in conjunction with other components to form larger units [2]: libraries, frameworks, web services, and entire applications. An important task of the architect of a software solution is to identify suitable software components – commercial, free or open source. To determine the suitability of components in a project context, software architects need to consider different aspects, such as license type, maturity, community support, etc.

As the number of components grows rapidly, a major problem in software reuse is the lack of efficient means to search and retrieve reusable components [9]. Usually, software architects rely on software components they know. Faced with new problem domains, they are left alone asking colleagues or consulting general-purpose search engines such as Google. Alternatively, they consult various sites for hosting software components such as GitHub, Apache.org or Sourceforge.net.

Wouldn't it be advantageous to use a semantic search application for software components? For example, by asking "Free Java library for machine learning?" a semantic search application would return a list of suitable products such as Weka and RapidMiner.

Semantic search applications have gained popularity in the last decade [8]. They have been established in a variety of application domains, including hotel portals, patent retrieval, dating sites etc. Surprisingly enough, semantic search applications in the computer science field, and in particular for software components , are not yet in widespread use: the shoemaker's son always goes barefoot.

This chapter presents the concept for a semantic search application for software components which we call "SoftwareFinder", as well as its implementation.

## 17.2   Example SoftwareFinder

We explain the user interaction concept of SoftwareFinder by means of the example of searching for "Free Java library for machine learning". SoftwareFinder offers a simple search box as known from search engines such as Google. A semantic auto-suggest feature offers suitable terms (see Fig. 17.1).

While the user is typing each letter of the search term (in this example "Machine Learning") into the box, the semantic auto-suggest feature offers suitable terms to complete the search query. Examples suggested are concrete products, such as "Torch5: fast *mach*ine learning toolbox", development terms, such as "Human *Mach*ine Interfaces" and general terms such as "*Mach*ine Learning". In addition to auto-suggest features as known, e.g., from Google, semantic auto-suggest offers other related terms with a semantic category, e.g., software product, programming language, business, development, general, etc. Additionally, synonyms and acronyms are considered.

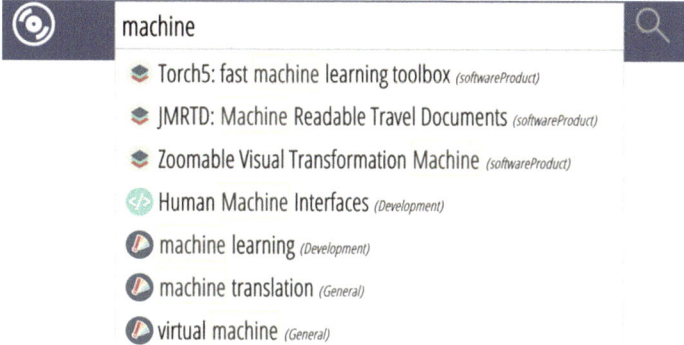

**Fig. 17.1** Semantic auto-suggest

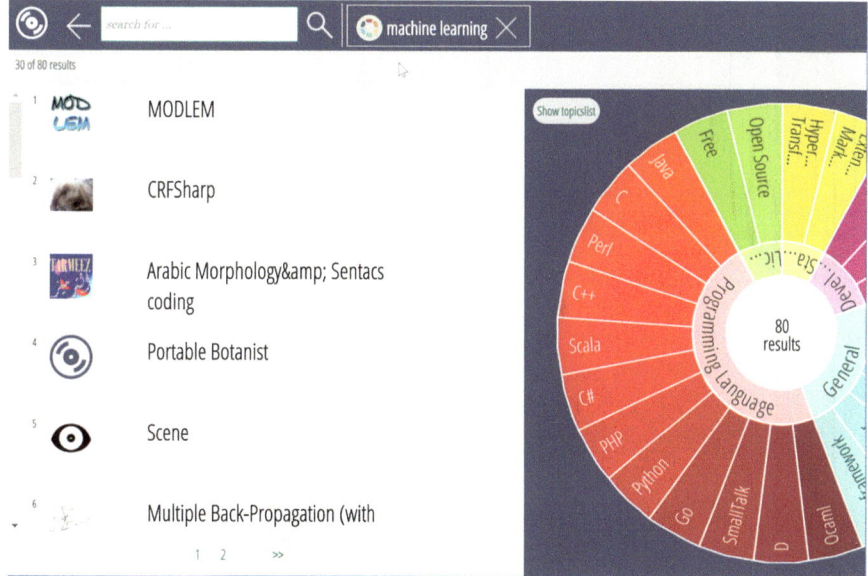

**Fig. 17.2** Topic pie

After entering the search term, a list of matching software components for the term "machine learning" will be displayed. Additionally, an innovative user interaction element which we call "topic pie" [1] supports a semantic faceted search (see Fig. 17.2).

The topic pie contains segments of distinct colour which denote different semantic categories as facets, e.g., programming language, licence, development etc. (inner ring). The outer ring contains concrete instances, e.g., programming languages Java, C#, Python, etc.

The user may now, iteratively, refine the search by selecting a slice of the topic pie, e.g., Java, Open source, etc. These sections will be displayed next to the search box (see Fig. 17.3).

In each refinement step, the list of search results as well as the topic pie is updated to match the current search criteria. Therefore, the number of results can be reduced from initially 150 (machine learning) to 35 (machine learning & Java) to finally 20 (machine learning & Java & open source).

The navigation (back button) allows returning to previous search states. Additionally, the search space can be explored in various dimensions: programming language, licence type, etc. For example, the programming language Java could be replaced by Python (see Fig. 17.4).

Finally, a product such as Weka may be selected and a detail page including a short description, tags, programming language, licence type and a star rating is displayed. Links to the project homepage and software download page are provided (see Fig. 17.5).

**Fig. 17.3** Stepwise refinement of semantic faceted search

**Fig. 17.4** Modifying the dimension "programming language"

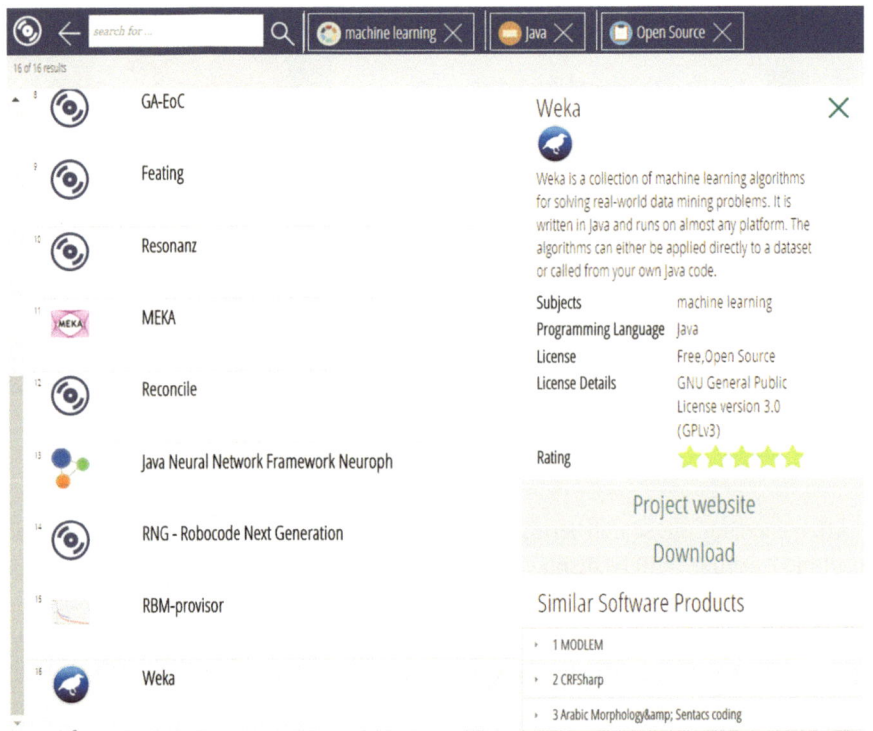

**Fig. 17.5** Software product details

**Fig. 17.6**  Similar software products

Similar Software Products

›  1 MODLEM

⌄  2 CRFSharp

## CRFSharp

CRFSharp(aka CRF#) is a .NET(C#) implementation of Conditional Random Fields, an machine learning algorithm for learning from labeled sequences of examples. It is widely used in Natural Language Process (NLP) tasks, for example: word breaker, postagging, named entity recognized, query chunking and so on.<br> <br> CRF#'s mainly algorithm is the same as CRF++ written by Taku Kudo. It encodes model parameters by L-BFGS. Moreover, it has many significant improvement than CRF++, such as totally parallel encoding, optimizing memory usage and so on. <br>

Semantic auto-suggest and semantic faceted search via topic pie are features which help the user to narrow down the search quickly. However, sometimes it is helpful to broaden the view, particularly when the user knows a particular product like "Weka" already but is interested in similar products. Recommendations like "Customers who bought this product also liked that one" are well-known from web shops. Software finder offers this functionality for recommending semantically similar software products (see Fig. 17.6).

SoftwareFinder is responsive. It runs on different devices such as a desktop computer, tablet computer and smartphone. The layout is automatically adapted to the screen size. On a smartphone, the topic pie is only partly visible at the left bottom corner of the screen. On touch, it will fully appear and can be used.

Responsive design is, of course, not a semantic feature. However, to improve the user experience of semantic applications, aspects like responsive design are as important as in other computer applications (see Fig. 17.7).

## 17.3    The Ontology as the Core of a Semantic Application

The ontology is the core of a semantic application such as SoftwareFinder. All semantic application logic such as semantic auto-suggest, semantic faceted search via topic pie, semantic search & ranking and recommendations are based on the ontology (see Fig. 17.8).

For SoftwareFinder, only a simple ontology, e.g. in the form of a thesaurus, is required for implementing the semantic application logic introduced above. It consists of terms

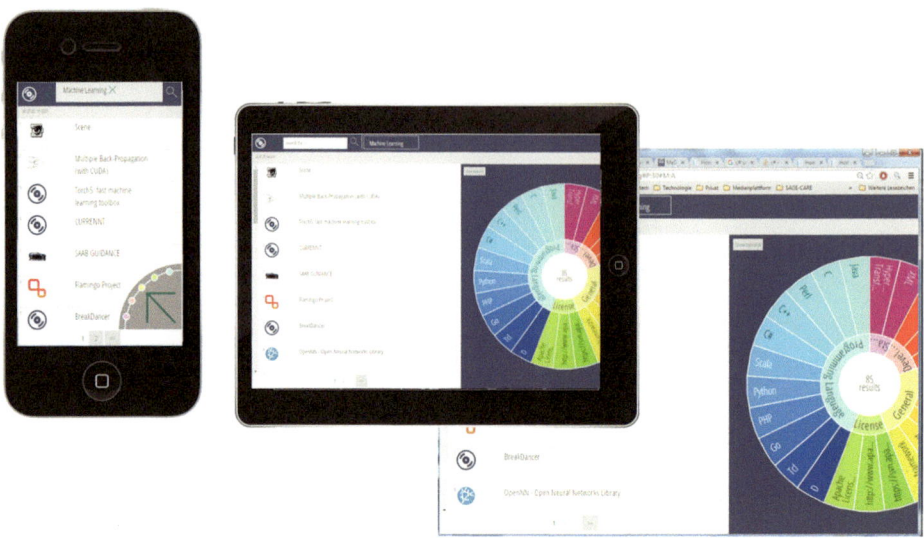

**Fig. 17.7** Responsive design [3]

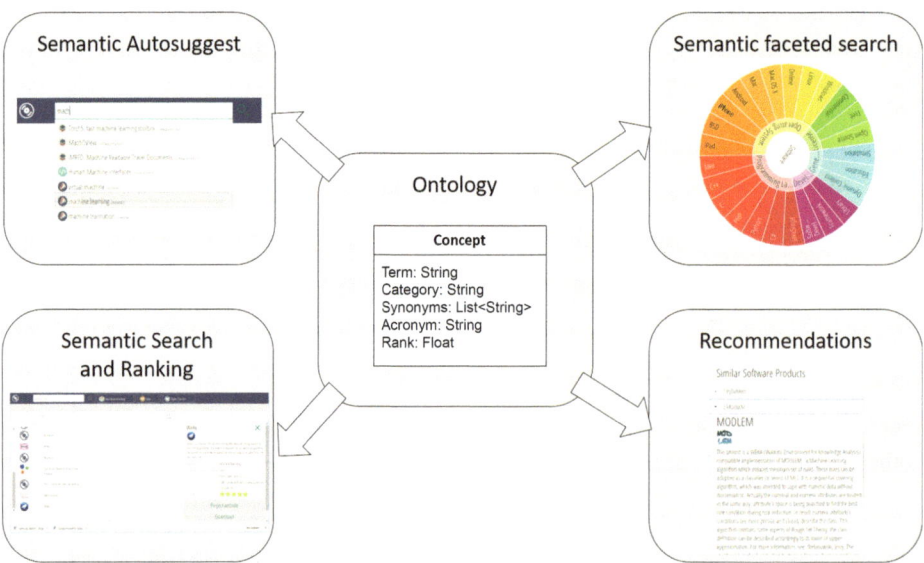

**Fig. 17.8** The ontology as the basis for semantic application logic

(e.g., "machine learning"), a semantic category (e.g., "general"), an acronym ("ML"), synonyms (e.g. "machineLearning" & "machine learnings"), and a rank (e.g., 0.6).

Concrete software products like Weka are described by ontology terms (tags, programming language, and license type), in addition to other attributes such as name, short description and download links.

### 17.3.1 How to Obtain the Ontology?

The ontology for SoftwareFinder, although simple in structure, needs to be sufficiently complete in order to be useful. This means that it must contain all the tags, programming languages and license types used of all software products that can be searched. All those terms must be semantically categorized and synonyms and acronyms should be provided where appropriate.

Best practice in developing semantic applications is to look for suitable off-the-shelf ontologies first. However, despite intensive search efforts, no off-the-shelf ontologies could be found which meets the requirements of SoftwareFinder. General-purpose ontologies such as DBpedia or YAGO are by far insufficiently specialised. Domain-specific classification systems and taxonomies such as from ACM or IEEE contain general terms and categories but no specific tags like "log analysis".

However, software hosting sites like "apache.org", "sourceforge.net", and "alternativeto.net" provide tags for describing the software components. Those tags can be used to derive the terminology behind the required ontology. Using crawling techniques, more than 20,000 terms could be identified. However, hand-crafting an ontology with more than 20,000 terms is time intensive. How is it best to proceed?

### 17.3.2 A Cost-Effective Methodology for Developing an Ontology from Large Domain Terminology

If no suitable off-the-shelf ontology is available for an application domain, a new ontology needs to be developed. A good starting point for developing an ontology is a set of terms which are relevant in the application domain. We call such a set of terms a *domain terminology*. However, if such a domain terminology is large, i.e., contains several thousand terms, then manually developing an ontology including all relevant terms is extremely costly.

For developing the SoftwareFinder ontology, we have successfully used a methodology which allows deriving an ontology from a large domain terminology in a semi-automated way [4]. The methodology is applicable in scenarios where the following conditions are met:

1. In a specific application domain, an ontology is needed and requirements for the ontology are specified.
2. A large terminology for the application domain, including several thousand terms, exists. These terms need to be included in the ontology.
3. There is no ontology available which meets the specified requirements.

See Fig. 17.9 for an overview of the methodology in BPMN Notation.

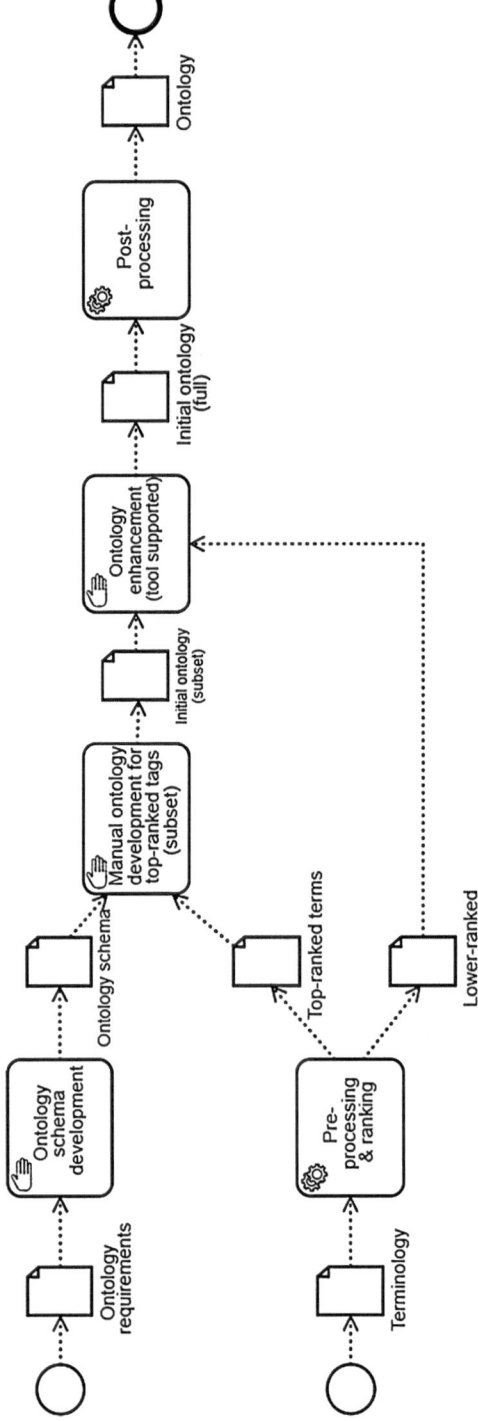

**Fig. 17.9**  Methodology for developing an ontology from a large domain terminology. (Adapted from [4])

### 17.3.2.1   Ontology Schema Development

Based on the ontology requirements, the domain-specific ontology schema needs to be developed. In the example of software component search, the ontology requirements are as follows. All terms for describing software components shall be assigned a semantic category, e.g., the category "business" for the term "enterprise resource planning" or the category "development" for the term "database management system". For "SoftwareFinder", 12 semantic categories have been identified. Acronyms shall be associated with terms, e.g., "DBMS" to "Database management system". Also, synonyms shall be associated with terms, e.g., "Database" with "Database management system".

Based on those requirements, the ontology schema consists of a single entity "Concept" which is shown in Fig. 17.10 as a UML class diagram.

### 17.3.2.2   Pre-processing and Ranking

In this step, the terminology is normalised and anomalies are handled in a domain-specific way. Furthermore, all terms are ranked according to relevance.

For SoftwareFinder, the following configuration data is used for normalization:

1. Blacklist (Ignore list): Some terms have no relevance for SoftwareFinder. For example, the term "Other/Nonlisted" is used in "sourceforge.net" which provides no useful information for the software component search. Therefore, such terms are specified in a blacklist which is used for removing them from the terminology as a pre-processing step.
2. Composite terms: Some terms are composites including several concepts. For example, the term "audio/video" indicates functionality for audio processing as well as for video processing. To improve searching, such terms can be split into multiple terms, e.g. "audio" and "video".

Ranking terms according to frequency is domain-specific. For SoftwareFinder, the number of software components that a term is assigned to is used as a heuristic for its relevance: the more often a term is used the more relevant it is considered.

The output of the step "pre-processing & ranking" is a list of normalised terms, sorted according to frequency ranking with the top-ranked terms first.

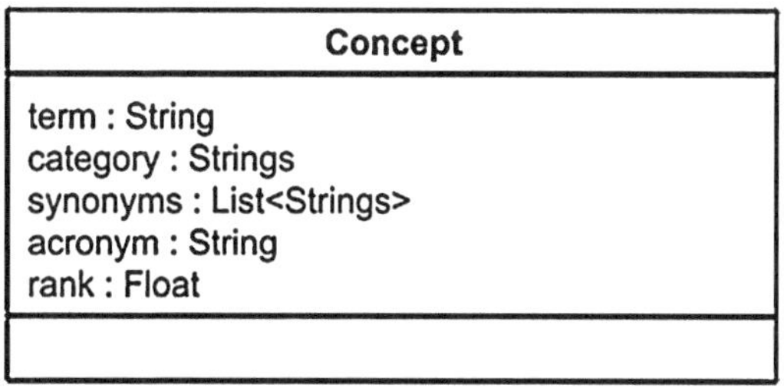

**Fig. 17.10** Concept entity
Class diagram [4]

| Concept |
| --- |
| term : String<br>category : Strings<br>synonyms : List<Strings><br>acronym : String<br>rank : Float |
| |

### 17.3.2.3   Manual Ontology Development

In this step, the domain expert takes a manageable subset of the terms, e.g., the first 500 top-ranked terms, and creates the corresponding concepts and relationships manually.

This includes manually assigning a term to a suitable semantic category. The output of this step is the first version of the ontology filled with a subset of the concepts containing the terms and their semantic categories.

### 17.3.2.4   Ontology Enhancement

In this step, the initial full ontology is developed, also including the lower-ranked terms. For this, domain-specific software tools may be used interactively.

In the example of software component search, simple software tools can be developed and used for identifying candidates for synonyms and acronyms. For example, containment of one term in another indicates a potential synonym, e.g., "word processor" is contained in "word processors".

Please note: linguistically speaking, "word processor" and "word processors" are not synonyms. However, we assume that users who manually tagged software components in hosting sites like sourceforge.net used both terms interchangeably.

Also, pattern matching techniques such as Jaro-Winkler [10] are used to find similarities which may indicate potential synonyms, e.g., "word processor" and "word processing". Using such tools, the domain expert can interactively enhance the ontology.

To find the potential acronyms among the terms within the ontology, the algorithm to extract abbreviations by Schwartz & Hearst [7] is used. The output of the algorithm is a list of all terms in the ontology with the suggested potential acronyms. The domain expert can go through the list manually and extend the ontology accordingly.

### 17.3.2.5   Post-processing

In this step, the ontology is finalized. For this, automated intelligent processes may be used.

In the example of software component search, the semantic categories for all lower-ranked terms (in total more than 20,000) need to be predicted. As input for prediction, the co-occurrence of terms in software components may be used. This is based on the assumption that terms, which are often used together for describing software components, belong to the same semantic category.

To verify this assumption, we have experimented with various supervised machine learning techniques [6]. The training set is based on the manually classified semantic categories of the top-ranked terms (about 700 terms). We used the data science platform RapidMiner (https://rapidminer.com/) and applied more than 10 machine learning techniques, including deep neural networks, Bayesian classifiers and decision trees. Using cross-validation [5], the overall accuracy was evaluated. However, the accuracy of all approaches ranged between 27% and 37% which is not considered sufficient.

To improve prediction accuracy, a domain-specific heuristic approach was implemented. For each term, all terms are collected which co-occur in any software product. The semantic category most often used in those co-occurring terms is used as a prediction of the semantic category. Consider the following examples:

1. "word processor" (expected category: "General"): General:259 (correct!), Development:38, Infrastructure:37, Business:12, Multimedia:8, Communication:7, Standards:6, Science:5, Humanities:5, Entertainment:3, Engineering:2, Security:1
2. "database management system" (expected category: "Infrastructure"): Infrastructure:81 (correct!), Development:43, General:31, Business:15, Standards:3, Science:1
3. "hypertext markup language" (HTML, expected category: "Standards"): Development:353 (incorrect!), General:147, Standards:63, Infrastructure:38, Communication:11, Business:4, Multimedia:2, Humanities:2

This approach offered an improvement in prediction accuracy over the machine learning approach. Out of 778 terms, 391 terms were correctly classified, 287 were incorrectly classified, and 109 could not be classified. The accuracy of predictions is 58%. A prediction is not possible for terms without classified co-occurring terms.

For SoftwareFinder, the prediction accuracy of the heuristic approach is sufficient since an incorrect semantic category is not mission critical. For example, the only effect of misclassifying HTML is that in the semantic auto-suggest and topic pie features, HTML will be displayed under the category "Development" instead of "Standards". The user of SoftwareFinder may be surprised by this but it will certainly not impede the semantic search.

## 17.4 How to Implement Semantic Functionality?

### 17.4.1 Software Architecture

Figure 17.11 gives an overview of the software architecture of SoftwareFinder [3]. The software architecture is separated into an online and an offline subsystem. The *offline subsystem* manages the crawling of the software hosting sites, regularly updating the SoftwareFinder data store as a batch process. The *online subsystem* performs the semantic search.

#### 17.4.1.1 The Offline Subsystem

The SoftwareFinder offline subsystem follows the software pipeline architectural style. A crawler visits software hosting sites and saves the HTML pages for the individual software components hosted. Afterwards, a semantic ETL (Extract, Transform, Load) process starts. It extracts metadata of the software components from the HTML pages, transfers them into a uniform format, preprocesses them semantically, and loads them into the data store.

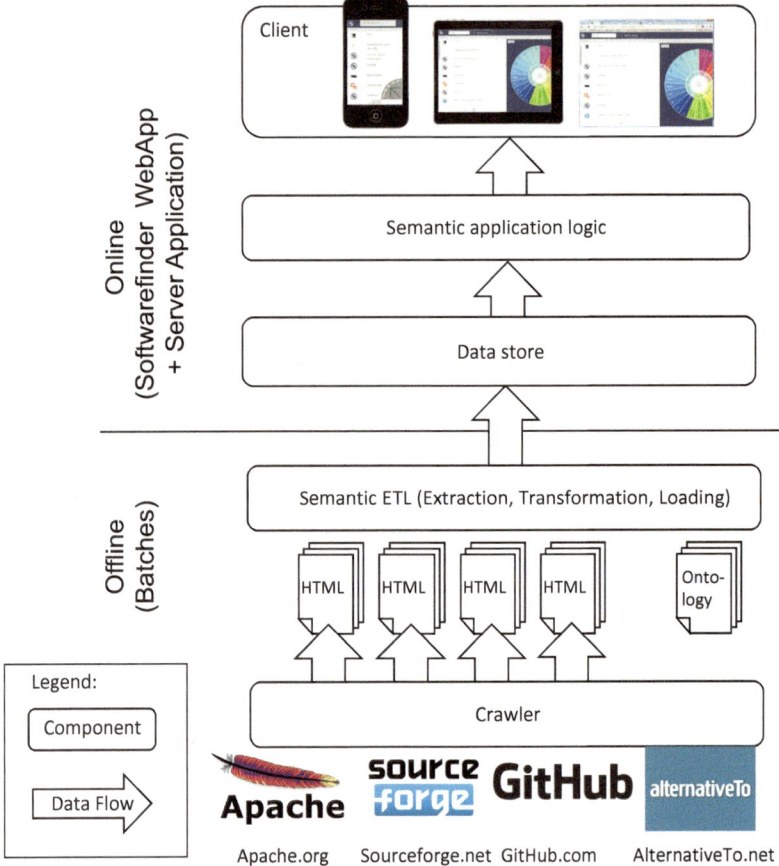

**Fig. 17.11** Software architecture [3]

### 17.4.1.2   Tag Normalisation

The ontology is used for automatically *normalising tags* of software components from different hosting sites (Fig. 17.12). For example, the synonym tags "Monitor" and "Monitors" are unified to the preferred tag "Monitor". Acronyms are handled, e.g., "CMS" is replaced by "Content Management System". Blacklisted tags such as "Other/Nonlisted Topic" are omitted and compound tags such as "Project and Site Management" are split up.

### 17.4.1.3   The Online Subsystem

The online subsystem is designed as a classical three-layer-architecture, consisting of client, semantic application logic, and data store. The data store contains metadata about software components and the ontology. It is indexed for high-performance access. The client implements the SoftwareFinder GUI including the responsive design. It accesses the semantic application logic via an API. The semantic application logic covers the various aspects: semantic auto-suggest, semantic faceted search, and recommendations.

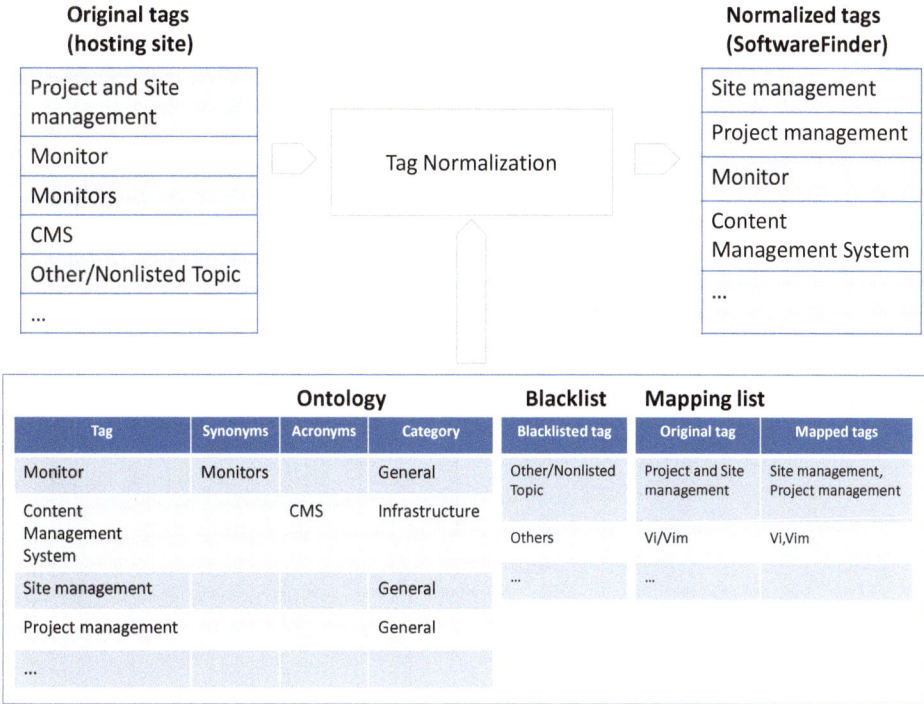

**Fig. 17.12**   Tag normalisation [3]

### 17.4.1.4   Semantic Auto-Suggest

For the semantic auto-suggest service, all concepts in the ontology are indexed, as well as all programming languages, operating systems, license types, and software product names. Initially, the tags matching a user input are ordered according to a heuristics-based relevance ranking. The heuristic used is as follows: the higher the term frequency, i.e., the more often a tag is used in the software product metadata, the higher its relevance.

Only the top 7 tags out of potentially hundreds of matching tags will be displayed to the user. Using relevance-based ranking only has a disadvantage: in many cases, tags of just one category will be displayed. In order to increase the category diversity, the initial, solely relevance-based ranking result is reordered. By omitting excess terms of the same category, it is ensured that the user has tags from at least three different categories to choose from, where relevant.

### Topic Pie Generation

The topic pie is generated in several steps (Fig. 17.13).

The input to the topic pie generator is the result set of the current search, i.e., a list of software product metadata. First, the tags are extracted from the metadata. The extracted tags are then grouped according to their categories and ordered using a heuristics-based relevance ranking. The heuristic used is: the higher the term frequency, i.e., the more often a tag is used in the result set, the more relevant it is. Therefore, the rank of an individual tag

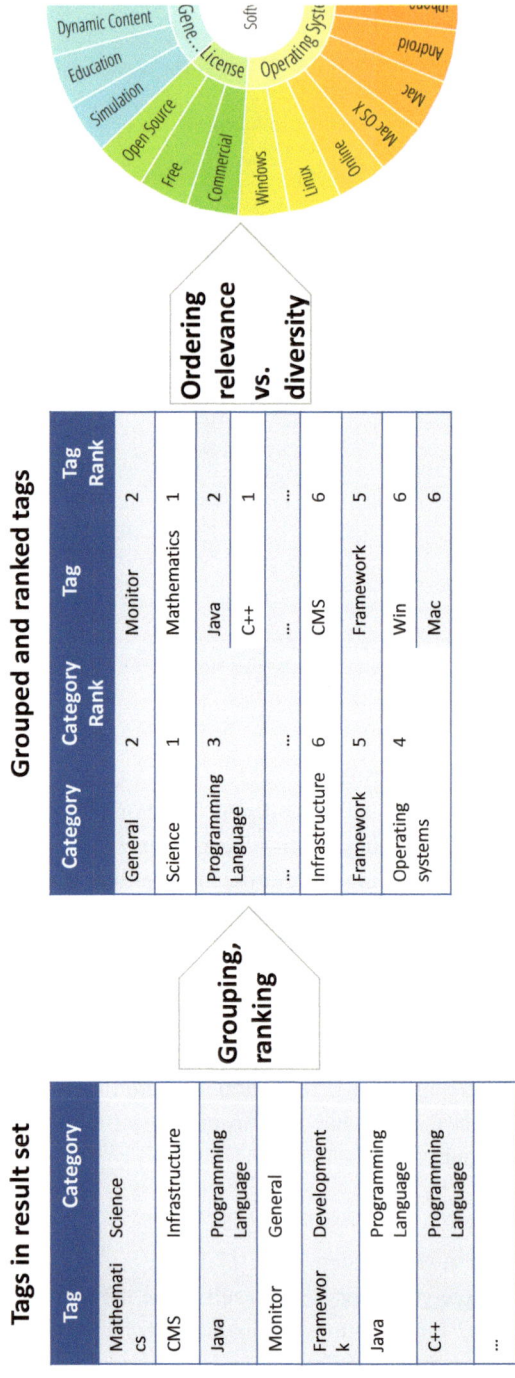

**Fig. 17.13** Topic pie generation [3]

is the number of its occurrences in the result set. The categories are ranked as well, based on the ranks of their individual tags: the more often a category is used in the result set, the more relevant it is. Therefore, the rank of a category is the sum of the ranks of its tags.

The topic pie accommodates up to 25 tags, out of potentially hundreds of selected tags. Selecting just the 25 top-ranked tags often results in only one or two categories being shown to the user. Therefore, as in the semantic auto-suggest feature, relevance is traded with diversity in order to display a well-balanced topic pie. By omitting excess tags of each category, it is ensured that the user has tags from potentially five different categories to choose from.

### Recommendations

For recommending software components which are similar to the one currently selected by the user, similarity metrics are used: the more tags two software component have in common, the more similar they are. Since all tags have been normalised during the semantic ETL process, issues of synonyms, acronyms, etc. need not be considered here. From the list of similar software components ordered according to the similarity metrics, the first three will be displayed to the user.

## 17.5  Implementation

We have successfully implemented SoftwareFinder. The server is implemented in Java 8 involving a number of third-party libraries: For crawling, the library crawl4j is used; the data store is implemented with Apache Lucene. Semantic search is implemented via Lucene's document fields. Ranking strategies are implemented using Lucene's custom boosting. Indexing for semantic auto-suggest uses Lucene's suggester based on infix matches called "AnalyzingInfixSuggester". Access performance even for complex queries is below 100 ms.

The client/server communication is via HTTP using JSON as the data format. The client web app is implemented in HTML5/CSS3/JavaScript using various JavaScript libraries: Knockout.js is used for implementing the MVVM architecture. JQuery, jQuery-UI and jQuery-touchSwipe are used for widgets and the client-server communication.

The server and the web app are deployed in an Apache Tomcat servlet container. SoftwareFinder can be used under the following URL: www.softwarefinder.org

## 17.6  Recommendations

We summarise our main learnings from implementing SoftwareFinder by means of recommendations.

1. In application domains where there is no suitable ontology available but a large set of domain vocabulary, semi-automatic ontology development may be a cost-effective approach. Manual classification work may be supported by smart tools, e.g., for suggesting synonyms and acronyms, as well as by heuristic methods and machine learning, e.g., for automatically classifying terminology.

2. When implementing a semantic search application consider providing the following features that are all based on the ontology:

    (a) Semantic auto-suggest: this enhances the auto-suggest feature known from search engines like Google with semantic information like category, synonyms, or acronyms.

    (b) Semantic faceted search via topic pie: options for refining the search (facets) are presented to the user based on the search results and the ontology. As an intuitive graphical widget, a topic pie can be used for displaying the semantic facets.

    (c) Recommendations: when selecting a search result, the user may be recommended alternative products which are semantically similar.

3. Search technologies such as Apache Lucene and Apache Solr are well suited for implementing Semantic search with high performance. Features like semantic auto-suggest, semantic faceted search via topic pie, and recommendations may well be implemented using search technology.

## 17.7 Conclusions

Semantic search applications may, indeed, improve the user experience. However, this does not come for free. A suitable ontology, development of semantic application logic with high performance as well as a carefully designed user interface are necessary.

We hope that SoftwareFinder is a good example of how Semantic search can be employed to improve user experience.

## References

1. Deuschel T, Greppmeier C, Humm BG, Stille W (2014) Semantically faceted navigation with topic pies. ACM Press, Leipzig
2. Hopkins J (2000) Component primer. Commun ACM 43(10):27–30
3. Humm BG, Ossanloo H (2016) A semantic search engine for software components. IADIS Press, Mannheim, pp 127–135
4. Humm BG, Ossanloo H (2017) Cost-effective semi-automatic ontology development from large domain terminology. In: Proceedings of the collaborative European research conference (CERC 2017), Karlsruhe
5. Larson SC (1931) The shrinkage of the coefficient of multiple correlation. J Educ Psychol 22:45
6. Russell SJ, Norvig P (2009) Artificial intelligence: a modern approach, 3rd edn. Pearson Education, Upper Saddle River
7. Schwartz AS, Hearst MA (2003) A simple algorithm for identifying abbreviation definitions in biomedical text. Pac Symp Biocomput 2003:451–462
8. Sirisha J, Subbarao B, Kavitha D (2015) A Cram on semantic web components. Int J Adv Res Comput Sci 6(3):62–67. 6 p
9. Spinellis D, Raptis K (2000) Component mining: a process and its pattern language. Inf Softw Technol 42(9):609–617
10. Winkler WE (1999) The state of record linkage and current research problems. US Census Bureau, Washington, DC

# Index

© Springer-Verlag GmbH Germany, part of Springer Nature 2018
T. Hoppe et al. (eds.), *Semantic Applications*,
https://doi.org/10.1007/978-3-662-55433-3